The corrosion performance of metals for the marine environment: a basic guide

European Federation of Corrosion and NACE International
Joint Publication
NUMBER 63

The corrosion performance of metals for the marine environment: a basic guide

Edited by
Carol Powell & Roger Francis

EUROPEAN FEDERATION OF CORROSION
FÉDÉRATION EUROPÉENNE DE LA CORROSION
EUROPÄISCHE FÖDERATION KORROSION

THE CORROSION SOCIETY

Published by CRC Press on behalf of the
European Federation of Corrosion,
The Institute of Materials, Minerals & Mining,
and NACE International

CRC Press
Taylor & Francis Group
Boca Raton London New York

CRC Press is an imprint of the
Taylor & Francis Group, an **informa** business

The Institute of Materials,
Minerals and Mining

First published 2012 by Maney Publishing

Published by CRC Press on behalf of the European Federation of Corrosion and The Institute of Materials, Minerals & Mining, and NACE International

2 Park Square, Milton Park, Abingdon, Oxon OX14 4RN
711 Third Avenue, New York, NY 10017, USA

CRC Press is an imprint of the Taylor & Francis Group, an informa business

ISBN 978-1-907975-58-5 (book)
ISSN 1354-5116
Cover image: Corroding steel piling on a UK beach. Image supplied courtesy of Roger Francis.

NACE International/EFC Joint Publication
The corrosion performance of metals for the marine environment: a basic guide

Item No. 24247

ISBN:978-1-907975-58-5
E-ISBN: 978-1-907975-59-2

European Federation of Corrosion
1 Carlton House Terrace
London SW1Y 5AF
United Kingdom
+44 20 7451 7336

NACE International
1440 South Creek Dr.
Houston, Texas 77084-4906
+1 281-228-6200

THE CORROSION SOCIETY

Contents

European Federation of Corrosion (EFC) publications: Series introduction

The European Federation of Corrosion (EFC), incorporated in Belgium, was founded in 1955 with the purpose of promoting European cooperation in the fields of research into corrosion and corrosion prevention.

Membership of the EFC is based upon participation by corrosion societies and committees in technical Working Parties. Member societies appoint delegates to Working Parties, whose membership is expanded by personal corresponding membership.

The activities of the Working Parties cover corrosion topics associated with inhibition, cathodic protection, education, reinforcement in concrete, microbial effects, hot gases and combustion products, environment-sensitive fracture, marine environments, refineries, surface science, physico-chemical methods of measurement, the nuclear industry, the automotive industry, the water industry, coatings, polymer materials, tribo-corrosion, archaeological objects and the oil and gas industry. Working Parties and Task Forces on other topics are established as required.

The Working Parties function in various ways, e.g. by preparing reports, organising symposia, conducting intensive courses and producing instructional material, including films. The activities of Working Parties are coordinated, through a Science and Technology Advisory Committee, by the Scientific Secretary. The administration of the EFC is handled by three Secretariats: DECHEMA e.V. in Germany, the Fédération Française pour les sciences de la Chimie (formely Société de Chimie Industrielle) in France, and The Institute of Materials, Minerals and Mining in the UK. These three Secretariats meet at the Board of Administrators of the EFC. There is an annual General Assembly at which delegates from all member societies meet to determine and approve EFC policy. News of EFC activities, forthcoming conferences, courses, etc., is published in a range of accredited corrosion and certain other journals throughout Europe. More detailed descriptions of activities are given in a Newsletter prepared by the Scientific Secretary.

The output of the EFC takes various forms. Papers on particular topics, e.g. reviews or results of experimental work, may be published in scientific and technical journals in one or more countries in Europe. Conference proceedings are often published by the organisation responsible for the conference.

In 1987 the, then, Institute of Metals was appointed as the official EFC publisher. Although the arrangement is non-exclusive and other routes for publication are still available, it is expected that the Working Parties of the EFC will use The Institute of Materials, Minerals and Mining for publication of reports, proceedings, etc., wherever possible.

The name of The Institute of Metals was changed to The Institute of Materials (IoM) on 1 January 1992 and to The Institute of Materials, Minerals and Mining with effect from 26 June 2002. The series is now published by CRC Press on behalf of The Institute of Materials, Minerals and Mining.

P. McIntyre
EFC Series Editor
The Institute of Materials, Minerals and Mining, London, UK

EFC Secretariats are located at:

Dr B. A. Rickinson
European Federation of Corrosion, The Institute of Materials, Minerals and Mining,
1 Carlton House Terrace, London SW1Y 5AF, UK

Mr M. Roche
Fédération Européenne de la Corrosion, Fédération Française pour les sciences de la
Chimie, 28 rue Saint-Dominique, F-75007 Paris, France

Dr W. Meier
Europäische Föderation Korrosion, DECHEMA e.V., Theodor-Heuss-Allee 25,
D-60486 Frankfurt-am-Main, Germany

Volumes in the EFC series

Preface

Many types of metallic material are used in a wide range of applications in marine environments. The materials' resistance against corrosion is vital for marine structural integrity and in order to avoid failure. In the 1970s, 1980s and 1990s, a lot of valuable knowledge was gained about how to avoid corrosion in marine and seawater service offshore. Much of this information was published in journals and conferences at the time, but it is not readily found electronically. This knowledge decreased the risk for corrosion problems. However, during recent years, it seems that corrosion problems are appearing again and that today's engineers are not aware of the lessons learned earlier.

In order to collect basic information together, the Marine Corrosion Working Party of EFC started to work on a book on the subject of corrosion in offshore and other marine environments, covering the common problems and how to avoid them. Early in the work, it was obvious that it would be beneficial to broaden the work with participants from NACE International as well. Therefore it is very natural to have this publication as a joint publication with NACE International. I would like to thank the members of both EFC WP9 and NACE STG 44 who wrote chapters or offered valuable comments on the early drafts.

It is our hope that the contents of the book will be useful to all scientists and engineers who are involved in the selection of materials for marine service.

Ulf Kivisäkk, *Chairman of the EFC WP9 on Marine Corrosion*

1

Introduction

Carol Powell

Vice Chair of EFC Working Party 9 for Marine Corrosion

carol.powell@btinternet.com

Seawater is not a simple solution of inorganic salts in water. It is a complex mixture containing many different salts, dissolved gases, trace elements, suspended solids, decomposed organic matter, and living organisms. Its composition can vary between wide limits; for example, the dissolved salt content of the Baltic Sea is about 7 g kg^{-1}, whereas the Persian Gulf has about 43 g kg^{-1}.

The temperatures of seawater range from 0 °C in the Antarctic to ~28 °C in the tropics [1]. Some shallow seas in equatorial regions can reach 40 °C, while 45 °C is typical of condenser outlet temperatures. Not only might seawater vary from location to location, but individual places can show changes in temperature and biological activity with the seasons.

As such, seawater can be an aggressive and unpredictable medium. For most metals and alloys, it is capable of causing degradation or *corrosion*. This can take the form of general thinning or more localised penetration in a variety of guises. The corrosion behaviour of metals in seawater is largely influenced by the oxygen content, seawater velocity, temperature, pollution, and marine organisms.

Organisms can occur on metals as microfouling (slimes) or macrofouling (e.g. sea grasses and mussels); both can have an influence on the corrosion behaviour of metals. Seawater is a good electrolyte and can potentially lead to galvanic corrosion of susceptible alloys if two or more are in electrical contact with each other and directly exposed to the seawater.

Alloys are mixtures of two or more different metals and many have the capability of improving the properties of the individual constituents. Service experience in seawater has developed over the years such that, with a reasonable understanding of the alloying additions and their effect, it is possible to have some estimation of their performance.

Corrosion of alloys is an electrochemical phenomenon and there is a trend to evaluate new alloy compositions in short-term laboratory trials rather than long-term live seawater exposures. Synthetic seawater standards are available for such evaluations, and the tests have their value, particularly in ranking similar alloys. However, the complex nature of seawater has made it difficult over the years to achieve realistic performance data from short-term testing, and it is still necessary to supplement this with long-term testing in natural seawater or service experience to fully understand how alloys perform.

The majority of alloys are susceptible to corrosion in seawater in one way or another. Their successful application relies on knowledge of the types of corrosion each alloy might be susceptible to and how to avoid potential problems by material selection, as well as good design, fabrication, and operational practices. Planned regular maintenance or replacement regimes can be a solution to limited service lives and be more economical when the system is accessible and labour is cheap.

Higher-reliability materials are the alternative solution and often provide lower life-cycle costs than the cheaper first cost, less reliable systems. In an ideal world, 'fit and forget' solutions would predominate, but no material is a panacea, and economic pressures are always present to prevent selection of more expensive choices unless absolutely necessary.

The goal of this guide is to provide readers with a basic understanding of the types of alloy used in seawater. All alloy groups discussed can form a successful part of engineering solutions for handling or withstanding seawater, but this is only achieved by designing and operating to their strengths and circumventing their weaknesses. The chapters include the types of corrosion problem areas encountered, and this knowledge can assist in avoiding them either at the outset or, if a failure has already occurred, in the replacement of the component.

The terminology used assumes the reader has no prior knowledge of corrosion engineering. The intent is to allow engineers new to the marine industry to quickly obtain an overview of important practical issues. However, it need not exclude more experienced engineers interested in expanding their breadth of knowledge about alloys in seawater, or refreshing their knowledge.

The chapters cover the main alloy groups starting with steel as the most widely used construction material. Each subsequent chapter discusses applications, the typical alloys used, basic properties, and corrosion resistance for the various alloy groups. The final chapter specifically examines galvanic corrosion because, of all the corrosion mechanisms, this is perhaps the easiest to misunderstand and regularly causes significant problems. The guide also is a reminder that few systems or structures are made of single alloys, and their success is ultimately linked to assessing the systems or structure as a whole, as well as the individual components.

Each chapter represents a short overview of the topics, and explanations are easy to follow. If further details are required, a reference and/or bibliography section is included at the end of each chapter. Photographs have been omitted to keep the guide as economical as possible; however, pictures of most of the types of corrosion described can be found in a case history publication by the European Federation of Corrosion [2].

References

1. R. Francis: 'The selection of materials for seawater cooling systems: A practical guide for engineers'; 2006, Houston, TX, NACE International.
2. Working Party on Marine Corrosion, European Federation of Corrosion (EFC): 'Illustrated case histories of marine corrosion', EFC Series Publication 5; 1991, Leeds, UK, Maney Publishing.

2

Iron and carbon steel

Roger Francis

Rolled Alloys, Unit 16, Walker Industrial Park, Walker Road, Blackburn BB1 2QE, UK

rfrancis@rolledalloys.com

2.1   Introduction

This chapter reviews the corrosion of carbon and low-alloy steels, as well as cast irons, in the uncoated condition. There are a number of coatings used with carbon steel to extend its life, but a discussion of these is outside the scope of this guide. Carbon and low-alloy steels are the most widely used materials in the marine environment, for both structural components and pressure-retaining applications. Table 2.1 lists some of the major applications.

Carbon manganese steels cover a range of compositions depending on the ductility and strength required. They are usually the lowest cost metallic materials. When additional properties are required (e.g. higher strength, low-temperature ductility), small amounts of nickel, chromium, molybdenum, and copper may be added, either individually or in combination, to create special low-alloy steels. This guide does not propose to list all of the alloys here because they corrode in more or less the same manner in seawater.

Cast irons contain 2–4% carbon and the microstructure contains graphite as well as carbides. Ductile iron is the most common alloy, and it contains small additions of magnesium to spheroidise the graphite and increase ductility. In many respects, cast irons corrode similarly to carbon steels in seawater.

Austenitic cast irons contain 15–25% nickel and small additions of other elements. ASTM (A571) [1] UNS F43010 (Grade D-2M) is often used in castings for pump cases and valve bodies, as well as for centrifugally cast pipe. It has a lower corrosion rate in seawater than ductile cast iron, but it has a susceptibility to chloride stress corrosion cracking (SCC) in warm seawater (see section 2.4.3 *Stress corrosion cracking* later in this chapter).

Table 2.1 Typical carbon steel and cast iron applications in marine environments

Type	Alloy	Application
Structural	Carbon steel and cast iron	Platforms, jackets, ship hulls, vessels, piling and sheeting, towers, bridges, lock gates, gratings, ladders, cranes, lifts, and loaders
Pressure-containing	Carbon steel, low-alloy steels, cast iron, and austenitic cast iron	Piping, vessels, pumps, and valves

All carbon steels and cast irons corrode in aerated seawater, so they are low-initial-cost alloys, but have high maintenance/replacement costs unless protective measures are used. Offshore maintenance is very expensive and carbon steel is chiefly used with high quality coatings and/or cathodic protection (CP). When maintenance is simple and planned shutdowns are frequent, carbon steel may be a cost-effective option.

2.2   Corrosion in the atmosphere

In marine atmospheres, the corrosion rate of carbon and low-alloy steels can vary from 0.01 to 0.1 mm year^{-1} depending on how close to the sea the exposed metal is located. In environments such as offshore platforms and ships' decks, where splashing is frequent, the corrosion rate is at the higher end of this range. Coupling to more noble (electropositive) metals can greatly increase the corrosion rate of carbon steel by galvanic corrosion at the junction to 1 mm year^{-1} or more. Corrosion of carbon steel leads to unsightly rust staining that may contaminate other equipment and cause accelerated attack. For example, corrosion products of carbon steel have caused localised pitting of UNS S31603 (316L) stainless steel (SS).

Low-alloy steels generally perform similarly to carbon manganese steel. Carbon steel with ~0.5% copper is termed *weathering steel,* and in rural and urban areas, it can acquire an adherent and protective orange-coloured film. However, in marine atmospheres, the copper has no protective effect; the corrosion rate is no lower than that of carbon steel and may be greater (in particular for non-rinsed applications) [2].

There have been some recent developments in low-alloy steels for marine atmospheres. It is claimed that additions of silicon and aluminium give improved corrosion resistance to steels used for structures such as marine bridges over harbours [3].

In most cases in which carbon steels and cast irons are used in marine atmospheres, high-quality coatings are needed for a useful, low maintenance life.

2.3   Splash zone

The intensity of corrosion of an unprotected steel structure in seawater varies markedly with position relative to the mean high and low tide levels (Fig. 2.1). Protection of a steel structure can be achieved by various means. Each corrosion zone shown in Fig. 2.1 must be considered separately. Generally accepted methods are CP, coating, and sheathing.

The spray and splash zone above the mean high-tide level is the most severely attacked region because of continuous contact with highly aerated seawater, and the erosive effects of spray, waves, and tidal actions. Corrosion rates as high as 0.9 mm year^{-1} at Cook Inlet, Alaska, and 1.4 mm year^{-1} in the Gulf of Mexico have been observed [4]. CP in these zones is ineffective because of the lack of continuous contact with the seawater (the electrolyte), and thus no current flows between the metals for much of the time. Coatings need to be very robust as the conditions can be damaging. Sheathing with neoprene, other rubber coatings, or nickel and copper alloys has been used successfully.

Corrosion rates for bare steel pilings, etc., can also be high at the position just below mean low tide, and are caused by galvanic effects because of the different levels of aeration that occur in the tidal region. This can be controlled by CP systems because the metal is continuously immersed.

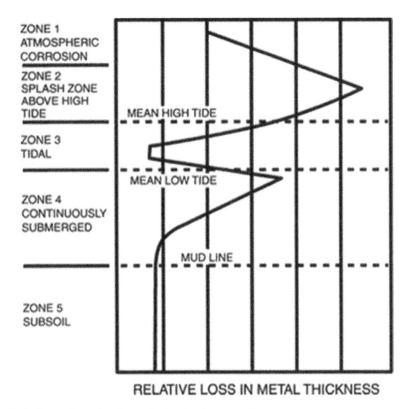

ZONE 1
ATMOSPHERIC
CORROSION

ZONE 2
SPLASH ZONE
ABOVE HIGH
TIDE

MEAN HIGH TIDE

ZONE 3
TIDAL

MEAN LOW TIDE

ZONE 4
CONTINUOUSLY
SUBMERGED

MUD LINE

ZONE 5
SUBSOIL

RELATIVE LOSS IN METAL THICKNESS

2.1 Corrosion of carbon steel piling in seawater

2.4   Immersion

2.4.1   General corrosion

In quiescent seawater, the general corrosion rate of carbon steel and cast iron is ~0.1 mm year^{-1}. However, cast iron and carbon steel can also undergo increased rates of corrosion under localised attack. General corrosion also increases as the seawater flow rate increases, as shown in Fig. 2.2 [5]. At a flow velocity of 4 m s^{-1}, the corrosion rate is approximately eight times that in quiescent seawater. Because austenitic cast iron contains nickel, it has a lower corrosion rate, typically 0.02 mm year^{-1} in quiescent seawater.

When steels and irons corrode in seawater, the corrosion rate gradually decreases because the corrosion products restrict diffusion of fresh water to the metal surface. For this reason, long-term corrosion tests (1 year or more) tend to yield lower corrosion rates than short-term tests.

When carbon steel is used as a sacrificial anode, fragmentation rather than general corrosion has sometimes been reported for high-carbon manganese and low-alloy steels. Mild steel (C < 0.1%) has been found to corrode more or less uniformly as an anode, although alloys with C < 0.15% are also reported as satisfactory.

Over time, the matrix of cast irons gradually corrodes, whereby the metallic constituents are selectively leached or converted to corrosion products, leaving the graphite particles intact, and the surface is slowly enriched in graphite (graphitic

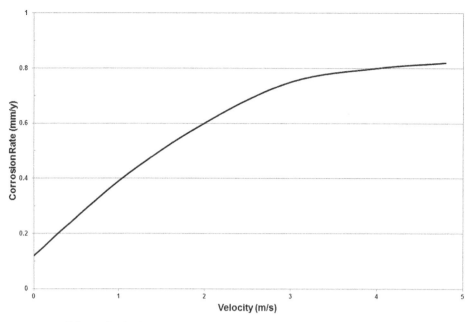

2.2 Effect of velocity on the corrosion of carbon steel in seawater [4,5]

corrosion). This causes the potential to become more noble (electropositive), and heavily corroded cast iron can become the cathode in a galvanic corrosion cell. This is a slow process in seawater and generally takes 10 to 20 years. It does not affect austenitic cast irons, because the corrosion rate is lower and components are not in service long enough for significant graphitic corrosion to occur.

Natural seawater (in contact with air) contains ~7 mg L^{-1} dissolved oxygen when saturated at approximately 20 °C, and this varies with temperature and depth [6]. In some cases, the seawater is deaerated, such as in multi-stage flash (MSF) desalination plants and seawater injection lines on offshore platforms. It is usual to express low levels of oxygen in parts per billion (ppb), where, for practical purposes, 1 ppb = 0.001 mg L^{-1}.

If the dissolved oxygen is maintained at ≤20 ppb, the corrosion of carbon steel is low. In most land-based plants, this can be achieved and maintained. Offshore, however, 50 ppb is a more common oxygen level and on some platforms, the control is so poor that levels of 100 to 500 ppb are not uncommon for extended time periods. Under these conditions, carbon steel develops significant corrosion.

The effect of dissolved oxygen concentration and temperature on the corrosion rate of carbon steel is shown in Fig. 2.3 [7]. These data assume that no protective films are formed. There is a sharp increase in the corrosion rate from 10 to 50 ppb O_2, and thereafter, a more gradual increase from 50 to 200 ppb O_2. The increase in corrosion rate from 10 to 80 °C at constant oxygen concentration is roughly the same order of magnitude as the increase produced by going from 10 to 50 ppb O_2 at 10 °C. The results show the importance of keeping the dissolved oxygen content under control when carbon steel is used.

The use of galvanised steel pipe is sometimes suggested as an alternative to bare carbon steel, because it is readily available and is superior to bare carbon steel in fresh

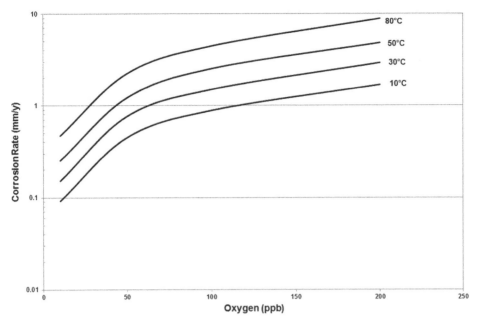

2.3 Effect of dissolved oxygen and temperature on the corrosion of carbon steel [7]

water. Galvanised steel offers no advantages in seawater because the zinc corrosion products are soluble. The zinc corrodes in a few weeks or months and then it acts just like bare carbon steel [8].

2.4.2   Pitting

Steels can undergo pitting to several times the depth of general corrosion under quiescent and flowing conditions. Hence, pitting corrosion is more likely to lead to leakage with carbon steel pipes [9]. The pit depth as a function of velocity is shown in Fig. 2.4. The effect of chlorine on the corrosion of carbon steel at various velocities is discussed later in section 2.4.7.

2.4.3   Stress corrosion cracking

Carbon steel and cast irons do not usually develop SCC in seawater. Although austenitic cast irons have a lower general corrosion rate in seawater (~0.02 mm year^{-1}) than grey cast iron, they are susceptible to chloride SCC, particularly in warm seawater.

Traditionally, austenitic cast iron components have lasted for at least 10 years, even in warm climates like the Middle East. In recent years, however, failures of pump cases have occurred within 1 to 6 years [10]. The shortest times to failure were on as-cast parts, while stress-relieved parts lasted a little longer. This reduction in life probably results from commercial pressures, because castings used to be grossly over-designed, but with increasing pressure to reduce costs, wall thicknesses are being reduced, with a consequent increase in the stress at critical locations.

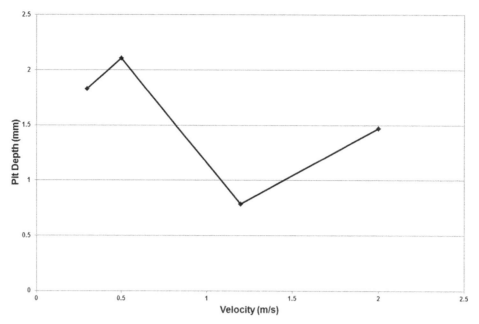

2.4 Effect of velocity on the pit depth of carbon steel after 14 months in seawater [9]

The number of failures in some countries, such as Saudi Arabia, has led to the use of more resistant materials, such as superduplex stainless steel. Note that in colder climates, such as the UK, where the seawater is typically 5 to 15 °C, pumps and valves in austenitic cast iron have lasted more than 20 years. However, coupling to a galvanically more noble metal can stimulate corrosion and could lead to SCC, even in cooler waters. Stresses in modern castings are usually higher than in the past, so the likelihood of SCC, even at moderate temperatures, is greater.

2.4.4   Hydrogen embrittlement

The normal potential of carbon and low-alloy steels in aerated seawater is about –550 mV SCE (saturated calomel electrode) and this is too electropositive to generate hydrogen. However, under the influence of CP, or coupling to more electronegative metals such as zinc, the potential can be –900 mV SCE or more negative, and hydrogen is then evolved. This does not cause a problem for most steels, but hydrogen embrittlement (HE) can occur with high-strength steels, such as those used for high-strength bolting or in jack-up rigs. This is avoided by using impressed current CP, low-voltage anodes, or Shottky diode limited conventional anodes, so the potential is about –850 mV SCE [11,12].

2.4.5   Fouling/microbiologically influenced corrosion

In natural seawater, carbon steel rapidly becomes covered with a thin fouling layer, followed quickly by attachment of shellfish and weeds. In some regions, this can happen very rapidly. In the Gulf of Mexico, a NPS 8 (200 mm nominal diameter) pipe became so full of fouling after 22 months that it had to be replaced [13]. Fouling of structures can also be severe enough to affect the mechanical loading.

When chlorine is added to the seawater to control fouling, this problem is eliminated, but there are consequent increases in pitting and general corrosion as described in section 2.4.7 *Chlorine/hypochlorite* later in this chapter.

When flow rates are high, approximately 2 m s^{-1} and above, there is less likelihood of attachment of fouling, but this varies according to different geographical locations, partly because of the differences in seawater temperature, and partly because of varying microbiological activity. At low flow velocities, suspended matter may settle onto the tube; as a result, the region beneath the deposits quickly becomes anaerobic. This stimulates the activity of organisms such as sulphate-reducing bacteria (SRB). These well-known causes for pitting corrosion of carbon steel may cause penetration of a pipe wall.

2.4.6   Accelerated low-water corrosion

The phenomenon of accelerated low-water corrosion (ALWC) was identified about 50 years ago and appears to be slowly spreading around the world. This is because there is a microbial involvement and it is thought that shipping may be spreading the bacteria involved in this form of attack [14].

ALWC affects sheet and tube piling and occurs between the mean low-water mark and the low astronomical tide mark. Corrosion rates can be 0.5 mm year^{-1} per side and because piling is normally installed uncoated, with a modest corrosion allowance, it is clear the life span of a piling can be drastically shortened.

The mechanism is thought to be a form of microbiologically influenced corrosion (MIC) and involves SRB becoming active beneath deposits and marine growths on the surface. The hydrogen sulphide can attack the carbon steel, but some can also diffuse into aerobic regions where sulphur-oxidising bacteria convert the hydrogen sulphide to sulphuric acid that further exacerbates attack of the steel. Although it is mostly seen on the seawater side of pilings, it can also occur on the landward side if there is sufficient seawater ingress and significant numbers of SRB [14].

A number of methods for combating ALWC have been tried, with the most successful being high quality coatings and CP. With piling, not all areas may be sufficiently protected by impressed current CP and a hybrid solution, combining both impressed current and sacrificial anodes, is required in some cases.

The whole phenomenon of ALWC, its causes and its prevention, has been reviewed and a document that includes an extensive bibliography was published by the International Navigation Association (PIANC) [14].

2.4.7   Chlorine/hypochlorite

Chlorine is added to seawater cooling systems to control macrofouling, such as that caused by weeds and shellfish. It also has the effect of keeping microfouling (slimes) at acceptable levels. The corrosion rate of carbon steel increases with chlorine concentration and velocity [9,15]. Figures 2.5 and 2.6 show the effect of chlorine on the general corrosion rate and pit depth, respectively, of carbon steel piping in seawater [9]. The exposure was for 14 months in the Gulf of Mexico with the seawater temperature varying from 12 to 29 °C.

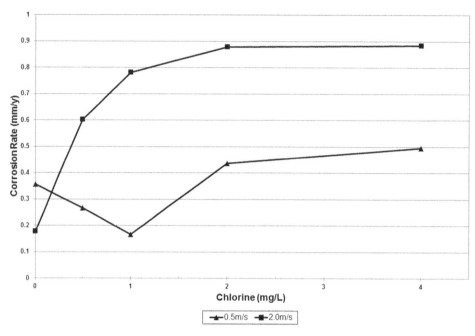

2.5 Effect of chlorine on the corrosion rate of carbon steel after 14 months in seawater [9]

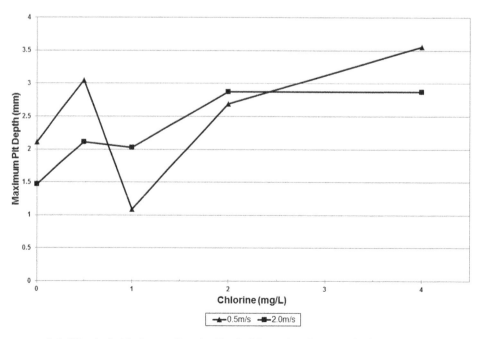

2.6 Effect of chlorine on the depth of pitting of carbon steel after 14 months in seawater [9]

The results show that pitting was deep and would lead to failure much more quickly than general corrosion. Figure 2.7 plots the safe continuous chlorine dose for no significant acceleration in corrosion vs. flow velocity using the maximum pit depth data [9]. This shows that, at 0.5 mg L^{-1} chlorine, the depth of pitting increases significantly above a flow velocity of 1 m s^{-1}. This design curve is not to eliminate pitting, but shows the chlorine/velocity level at which corrosion is no worse than without chlorine.

2.4.8   Galvanic corrosion

When cast iron and carbon steel are coupled to most metals, the corrosion rate increases, whereas the level of corrosion decreases for the coupled metals. This is because of galvanic effects and the cast iron or steel becomes the anode (see Chapter 8 on *Galvanic corrosion*). Because of their low cost, cast iron and carbon steel are frequently used as sacrificial anodes, for example, in the protection of brass tube plates in heat exchangers, and are purposely applied to corrode and decrease, or eliminate, the corrosion of the more vital component.

When cast iron corrodes, the graphite is left behind. Over a period of time, this means that the cast iron surface becomes soft and the component is much weaker. In addition, the graphite surface is highly cathodic to most metals, which can lead to accelerated galvanic corrosion of these metals.

2.4.9   Fabrication

One area that is often neglected when materials selection is considered is that of welds. This is one of the most common methods of joining, and the composition and structure of the weld metal may be quite different from that of the parent metal [16].

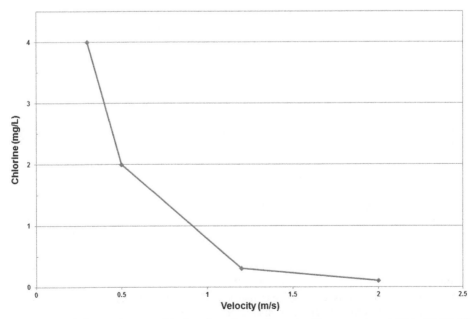

2.7 Safe continuous chlorine dose for carbon steel in seawater at 12 to 29 °C [9]

It is important, therefore, that the weld metal is not galvanically more susceptible to corrosion than the parent metal. Because the weld metal area is relatively small compared with the adjacent parent metal, any attack at the weld is greatly accelerated (see Chapter 8, *Galvanic corrosion*).

With carbon steel, problems of preferential corrosion of the weld metal usually result from the choice of welding consumables. High-silicon, low-nickel consumables give welds that are rapidly corroded in seawater, while the reverse (i.e. low-silicon, high-nickel) can produce welds that are acceptable. A common filler used to avoid preferential weld corrosion contains 0.6 to 1.0% nickel and 0.4% copper [16].

Another problem that sometimes occurs with carbon steel is preferential corrosion of the heat-affected zone (HAZ). If the welding is carried out with a heat input that is too low, then the HAZ can be hard and develop preferential corrosion in some aqueous media at room temperature. The solution is to weld with a higher heat input to produce a softer HAZ. This can be a problem with some alloyed steels, because welding produces a more coarse grain size than occurs in the parent metal [17].

Bond presented a paper reviewing the corrosion of carbon and low-alloy steel welds, the causes of problems, and the methods of avoidance [16]. Also described are potential problems with underwater weld repair and how to minimise them.

Acknowledgements

The author would like to acknowledge the assistance given by Stuart Bond in the writing of this chapter.

References

1. ASTM: 'Standard specification for austenitic ductile iron castings for pressure-containing parts suitable for low-temperature service', ASTM A571/A571M – 01(2011), ASTM, West Conshohocken, PA, 2011.
2. V. Kucera and E. Mattsson: in 'Atmospheric corrosion', (ed. W. H. Ailor), 1982, New York, NY, John Wiley & Sons.
3. T. Nishimura: 'Atmospheric corrosion of Si and Al-bearing ultrafine-grained weathering steel,' EUROCORR 2007, Freiburg, Germany, September 2007, European Federation of Corrosion.
4. R. W. Ross, Jr and D. B. Anderson: Proc. '4th Int. Congress on Marine corrosion and fouling', 461–473, Antibes, France, 1976.
5. INCO: 'A guide to the selection of marine materials', 2nd edn.; 1973, Toronto, Ontario, Canada, Nickel Institute.
6. H. U. Sverdrup, M. W. Johnson and R. H. Fleming: 'The oceans, their physics, chemistry, and general biology'; 1942, New York, NY, Prentice-Hall.
7. J. W. Oldfield and B. Todd: *Desalination,* 1979, **31**, (1–3), 365.
8. P. Ffield: *J Soc. Nav. Eng.,* 1945, **57**, (1), 1–20.
9. V. B. Volkening: *Corrosion,* 1950, **6**, (4) 123.
10. R. Francis: 'The selection of materials for seawater cooling systems: A practical guide for engineers'; 2006, Houston, TX, NACE International.
11. J. P. Pautasso, H. Le Guyader and V. Debout: CORROSION/98, San Diego, CA, March 1998, NACE International, Houston, TX, Paper no. 98725.
12. E. Lemieux, K. E. Lucas, E. A. Hogan and A.-M. Grolleau: CORROSION/2002, Denver, CO, April 2002, NACE International, Houston, TX, Paper no. 02016.
13. E. K. Albaugh: *World Oil,* 1984, **199**, (6), 94.

14. Working Group 44: 'Accelerated low water corrosion,' A report of the Maritime Navigation Commission, PIANC, Brussels, Belgium, 2005.
15. D. B. Anderson and B. R. Richards: *J Eng. Power,* 1966, **88**, 203–208.
16. S. Bond: in 'Prevention and management of marine corrosion', London, 2–3 April 2003. Lloyds List Event. TWI, Cambridge, UK, 2003. Organiser: Lloyds Register.
17. R. Francis: 'Galvanic corrosion: A practical guide for engineers'; 2000, Houston, TX, NACE International.

Suggested further reading

R. Covert et al.: 'Properties and applications of Ni-resist and ductile Ni-resist alloys.' Technical Series 11018, Nickel Institute Publications, Toronto, Ontario, Canada, 1998.
F. L. LaQue: 'Marine corrosion: Causes and prevention'. Corrosion monograph series; 1975, Hoboken, NJ, John Wiley & Sons.
T. H. Rogers: in 'Newnes international monographs on corrosion science and technology', 307; 1968, London, UK, George Newnes.
M. Schumacher, ed.: 'Seawater corrosion handbook'; 1979, Park Ridge, NJ, Noyes Data Corp.

3
Stainless steels

Roger Francis

Rolled Alloys, Unit 16, Walker Industrial Park, Walker Road, Blackburn BB1 2QE, UK
rfrancis@rolledalloys.com

3.1 Introduction

Stainless steels provide a wide range of strengths and corrosion resistance in seawater environments. Some have limitations that restrict their use to marine atmospheres or require cathodic protection (CP). Others have been developed to a sophisticated level and have extremely high corrosion resistance in seawater. The degree of alloying, temperature, seawater flow, oxygen levels, chlorination, and welding considerations can all influence the performance of stainless steels, and are considered in relation to alloy selection and application.

3.2 Characteristics of stainless steels

Stainless steels have negligible general thinning in seawater because of a protective, predominantly chromium oxide film that forms immediately on their surface with exposure to air. Oxygen levels present in seawater normally allow the film to repair itself if damaged; in fact, oxygen levels as low as 10 ppb, such as those found in thermal desalination plants, are still sufficient for this to occur.

The film is maintained at very high flow rates, and seawater velocities in excess of 40 m s^{-1} can be accommodated [1]. In practice, however, flow rates in offshore stainless steel pipe systems are often limited to 7 to 10 m s^{-1} in stainless steels as a result of pumping costs and noise restrictions. Even so, the combination of good strength and erosion resistance reduces weight and is economical.

Good manufacturing and fabrication practices are paramount in obtaining the best performance from stainless steel, which is readily fabricated and welded [2–6]. Even the strong duplex stainless steels exhibit good ductility, and there are many suppliers and fabrication shops with experience working with these metals.

Under certain conditions in environments containing oxygenated chloride, the protective surface film on stainless steels can break down locally, leading to pitting, crevice corrosion, or chloride stress corrosion cracking (SCC). Alloy grades can be selected with increased additions of chromium, molybdenum, and nitrogen to significantly improve resistance to crevice corrosion and pitting. Higher nickel levels, or partial or full ferritic structures, increase resistance to chloride SCC compared with 300 series austenitic alloys. Therefore, the situation often faced during material selection is identifying the type of stainless steel that has the correct corrosion resistance required for any given marine environment.

3.3   Types of stainless steel in marine use

3.3.1   Applications

Stainless steels are used for a wide range of applications in seawater and for many different reasons. Corrosion resistance in seawater is often only one factor; others include its strength, fabricability, flow velocity, weight saving, and corrosion resistance to other conditions in addition to seawater. Table 3.1 provides examples of how stainless steels are used in marine environments. They may be part of a mixed metal system and be protected by other less noble alloys, in which case a high corrosion resistance is not the prime requirement.

Table 3.1 Typical stainless steel applications in marine environments

Conditions	Alloy type	Applications
Marine atmospheres	UNS S31600 (316)/ UNS S31603 (316L); Superduplex	Instrumentation tubing, electrical connectors, instrument housings, rebar in concrete, handrails, cable trays, platform module cladding, fasteners, and boat hardware
Seawater with galvanic protection	UNS S31600 (316)/ UNS S31603 (316L); UNS S32205 (2205)	Tube sheets, hull-mounted equipment, pump impellers, valves, stems and trim, fasteners for Al and steel, and pump shafts
	Lean duplex Duplex	Thermal desalination plants Subsea flow lines handling wet CO_2 and instrumentation tubing
	Precipitation-hardened grades	Special fasteners
Seawater without galvanic protection	UNS N08904 (904L)	Pipes, trays, and spray heads in thermal desalination plants
	6% Mo	Power plant condensers, condenser tubing, firewater, ballast and seawater pipes, pumps, valves, and reverse osmosis (RO) desalination
	Superduplex	High-pressure oilfield injection pumps, seawater and firewater pumps, shafting, propellers, retractable bow plane systems, seawater piping and valves, fasteners, heat exchangers (tubes and plates), umbilicals, and RO desalination piping
	Superferritics	Power station condenser tubing
Deaerated brines	UNS S31600 (316)/ UNS S31603 (316L); Duplex	Desalination flash chamber lining and water injection tubing

Alternatively, they may be required to perform on their own merits when attention to corrosion resistance and design is paramount. More details about the range of stainless steels commonly available and their properties can be found in reference 7. Table 3.2 and the subsequent paragraphs provide details of the grades of stainless steels in marine service, while Table 3.3 shows the minimum mechanical properties at room temperature.

3.3.2   Austenitic stainless steels

Austenitic stainless steels have a tough, ductile structure and are the most commonly available and versatile type of stainless steel. A selection of common austenitic grades is included in Table 3.2. For welded components, only low-carbon (<0.03%) or stabilised grades should be used, to ensure freedom from intergranular corrosion in the heat-affected zones. The conventional basic austenitic grade used in seawater is the UNS S31600 (316) alloy, which contains 17% Cr, 10% Ni, and 2% Mo; the L grade (UNS S31603 [316L]) contains low carbon. These have a limited resistance to localised corrosion and, in the presence of crevices or under quiet exposure conditions, require galvanic protection from surrounding components or CP. It should be noted that today UNS S31600 (316) and UNS S31603 (316L) are often produced to the minimum chemical composition that has lower corrosion resistance in a marine atmosphere than the more alloyed version used in the past [8,9].

A more highly alloyed austenitic alloy is UNS N08904 (904L), with higher levels of chromium and molybdenum than UNS S31603 (316L). Originally, it was developed for sulphuric acid service, but has seen some use in seawater, particularly desalination.

Increased alloying additions of chromium, molybdenum, and nitrogen can achieve significantly higher resistance to localised corrosion. Over the past 20 years, a number of proprietary 5–7% molybdenum alloys have been developed, with high levels of nitrogen that have the added benefit of stabilising the austenite and enabling the alloys to be produced in thick sections, as shown in Table 3.2. These are known as super austenitics or 6% Mo alloys.

3.3.3   Duplex stainless steels

Duplex stainless steels contain both austenite and ferrite in their structure in roughly 50:50 proportions. They have higher chromium but lower nickel content than the austenitic alloys and low carbon content, as shown in Table 3.2. The lean duplex stainless steels contain low nickel and molybdenum contents that make them competitive with UNS S31603 (316L) (see Table 3.2). The lower-molybdenum-containing alloys have a localised corrosion resistance similar to or slightly inferior to that of UNS S31603 (316L). The higher molybdenum lean alloys are somewhat superior to UNS S31603 (316L) in terms of resistance to localised corrosion.

UNS S32205/S31803 (alloy 2205) with 22% chromium has better localised corrosion resistance than UNS S31600 (316) grades, and all of the duplex alloys have improved resistance to chloride stress corrosion cracking (SCC). Yield strength for the lean duplex alloys and UNS S32205 (2205) is about double that of the austenitics; duplex alloys have a strength of 450 MPa compared with 190 MPa for UNS S31603 (316L). Higher (~25%) chromium, molybdenum, and nitrogen-containing superduplex grades, sometimes with copper and tungsten, are also available both in wrought and

Table 3.2 Typical nominal compositions for some common wrought stainless steels

Type	Common name	UNS No.	Nominal composition (wt%)								PREN[a]
			C	Cr	Ni	Mo	N	Cu	W	Other	
Austenitic	316	S31600	≥0.08	17	10	–	–	–	–		17
	316L	S31603	≤0.03	17	10	2	–	–	–		24
	904L	N08904	≤0.03	20	25	4	–	1.5	–		34
	6% Mo	S31254	≤0.03	20	18	6	0.2	0.7			43
	6% Mo	N08926	≤0.03	20	25	6	0.2	0.7			43
	6% Mo	N08367	≤0.03	20	24	6	0.2	–			43
Duplex	Lean duplex	S32101	≤0.03	21	1	–	0.15	–	–	Mn	24
	Lean duplex	S32003	≤0.03	20	3	2	0.15	–	–	Mn	>30
	Duplex	S31803/ S32205	≤0.03	22	5	3	0.15	–	–		35
	Superduplex	S32760	≤0.03	25	7	3.5	0.25	0.7	0.7		>41
	Superduplex	S32750	≤0.03	25	7	3.5	0.25	–	–		>41
Ferritic	Superferritic	S44735	≤0.03	29	0.5	4	0.045	–	–	Nb	42
		S44660	≤0.03	28	1	3.5	<0.04	–	–	Ti, Nb	40
Precipitation-hardening	17-4 PH	S17400	≤0.07	17	4	–	–	4	–	Nb	17
	Grade 660	S66286	≤0.08	14.5	25	1	–	–	–	Ti, V, B	18

[a]PREN (pitting resistance equivalent number) is calculated using Equation 3.1.

Table 3.3 Minimum mechanical properties of wrought alloys at room temperature

Type	Common name	UNS No.	0.2% Proof stress (MPa)	Tensile strength (MPa)	Elongation (%)	Hardness[a] (HRC)
Austenitic	316	S31600	213	500	45	22
	316L	S31603	213	500	45	22
	904L	N08904	230	530	40	22
	6% Mo	S31254	300	650	35	22
	6% Mo	N08925	300	650	35	22
	6% Mo	N08367	300	650	35	22
Duplex	Lean duplex	S32101	450	650	25	28
	Lean duplex	S32003	450	650	25	28
	Duplex	S31803/ S32205	450	650	25	28
	Superduplex	S32760	550	750	25	28
	Superduplex	S32750	550	750	25	28
Ferritic	Superferritic	S44735	415	520	18	25
Precipitation-hardening	17-4 PH	S17400	750–1450[b]	1030–1340[b]	5–12[b]	33–45[b]
	Grade 660	S66286	585–650[b]	895–1000[b]	5–15[b]	24–37[b]

[a]Maximum.
[b]Depending on heat treatment.

cast forms and these combine higher strength with the high corrosion resistance of the superaustenitic alloys (Tables 3.2 and 3.3).

3.4   Precipitation-hardened, martensitic, and ferritic stainless steels

Although austenitic and duplex stainless steels are the most commonly used in marine service, other grades, including ferritic and martensitic stainless steels, are available. While ferritic and martensitic stainless steels do not possess the same levels of corrosion resistance as austenitic and duplex alloys in marine environments, most have much better resistance to atmospheric marine corrosion than carbon and low-alloy steels. Generally they are not recommended for immersed conditions.

Superferritic alloys with high levels of chromium and other alloying additions, such as molybdenum and nickel (Table 3.2), show good resistance to corrosion both in marine atmospheric and immersion service [10]. However, they are difficult to manufacture in thick sections, and their use is generally restricted to thin sections to avoid toughness problems. They are mostly used as thin-walled heat exchanger tubes in power station condensers.

Precipitation-hardened grades have chromium and nickel as the primary alloying elements. They also contain elements such as Al, Cu, Nb, Ti, and Mo. They are principally used because of their high strength, but are seldom used at their highest strength levels because of susceptibility to hydrogen embrittlement (HE) and SCC. Their corrosion resistance is not as good as many austenitic and duplex stainless alloys, so they are only used when strength is the prime requisite, and corrosion-resistance requirements are moderate, or when they are galvanically protected by lower-alloy materials, such as carbon steel.

3.5   Corrosion resistance

3.5.1   Marine atmosphere

As chromium is added to carbon steel, the corrosion rate in a marine atmosphere decreases until at 12–14% chromium, the rate becomes difficult to measure. At 18% chromium and above, the weight loss is negligible. Ferritic stainless steels tend to show a surface rusting and shallow pitting that has little effect on section thickness but is not always acceptable aesthetically. Standard austenitic grades, for example, UNS S31603 (316L), can retain their bright appearance much longer, particularly when periodic washing down removes surface-deposit build-up, and superficial staining [11,12]. Higher-alloyed stainless steels also retain their bright appearance under more severe conditions.

The severity of marine atmospheric corrosion depends on several factors, notably the degree of salt deposition, which correlates with the direction of prevailing winds and distance from the surf, and the humidity. Because sea salt contains several constituents of varying hygroscopic tendencies, brine is present on the surface even under what would be considered 'dry conditions' – relative humidity as low as 50% at room temperature. Rainfall or cleaning can wash away the salt deposits and reduce the effect of marine corrosion. In structures, location and orientation strongly influence the extent of marine atmospheric corrosion, as shown in Fig. 3.1.

Pitting and crevice corrosion can occur on UNS S31603 (316L) when used for hydraulic and instrumentation tubing. This was initially reported in the 1970s but its occurrence has increased during recent years. The problem is not limited to hot climates even though it is more common in subtropical and tropical areas. Field testing shows that 6% Mo and superduplex are successfully used in marine

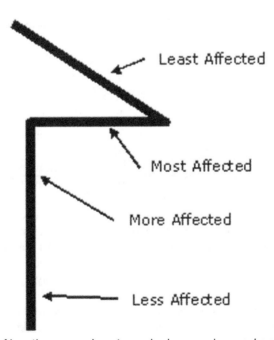

3.1 Effect of location on marine atmospheric corrosion – schematic diagram of roof overhang

atmospheres as hydraulic and instrument tubing if the clamps are properly installed [9]. Polymeric sheathing is another option that has been used with good results [13].

For stainless steel subject to internal elevated temperatures, SCC can occur in the marine atmosphere and the following typical temperature limits have been applied: 50 to 60 °C for UNS S31600 (316); 100 to 120 °C for 6% Mo; 80 to 100 °C for duplex; and 90 to 110 °C for superduplex. It should be noted that the lower temperature limits are based on aggressive laboratory testing while the higher limits have been applied by end users without problems [14].

3.5.2   Fully immersed pitting and crevice corrosion

Pitting of boldly exposed (non-creviced) base metal is an indication there are surface inclusions or imperfections in the metal, chlorination level is too high (as discussed later in this chapter) or the environment is too aggressive for the stainless steel grade, and a more resistant grade may be required. Pitting of stainless steels is usually seen in welds used beyond their service limits, or in the weld/heat-affected zone of poorly fabricated joints.

Typical areas for crevice corrosion are found beneath O-rings, flange faces under gaskets, threaded connections, nonmetallic connectors, tube-to-tube-sheet rolled joints, under adhesive tape, and under oxide scale. The tighter and deeper the crevice, the more severe it is. Such corrosion can also be found in quiet seawater conditions, normally at velocities less than 1 m s^{-1} when sediment or marine fouling is allowed to deposit.

Although the mechanisms for pitting and crevice corrosion are slightly different, they both involve localised oxygen depletion accompanied by metal dissolution, migration of chloride ions, and lowering of the local crevice pH. Crevice corrosion occurs more readily than pitting in seawater and at lower temperatures in the same bulk environment. Higher levels of the same alloying additions improve resistance to both types of corrosion.

Useful ranking of alloys is carried out by the pitting resistance equivalent number (PREN). See Equation 3.1, or critical pitting or crevice temperature tests in ASTM G48 [15].

$$PREN = \%Cr + 3.3\,(\%Mo + 0.5 \times \%W) + 16 \times \%N \qquad [3.1]$$

The higher the values, the better the resistance to pitting and crevice corrosion. A PREN in excess of 40 is normally considered necessary for an alloy to be considered for aerated seawater [16,17]. Alloys with a low PREN number, for example, alloys UNS S31603 (316L) and UNS S32205 (2205), usually require CP if not galvanically protected by other metals in the component or system [18].

When metal and environmental conditions are such that crevice corrosion is a possibility, attention should be paid to designing out crevices, for example, by avoiding threaded connections, sealing tight crevices with welds, and making full-penetration pipe welds. Alternatively, local improvements to crevice corrosion resistance are achieved at areas such as flange faces by weld overlaying with more resistant high-nickel alloys [19].

Biofilms form very quickly in seawater on metal surfaces and tend to give the stainless steel a more noble corrosion potential and increase the likelihood of initiation, and also the propagation of pitting and crevice corrosion compared with sterile water. Biofilm activity ceases at a temperature of 25 to 30 °C above normal ambient temperature [20], with the result that corrosion attack can be higher in ambient-temperature water than if the same water is heated to a temperature above which the bioactivity decreases significantly [21].

Chlorination, in optimal doses, is favourable to the corrosion performance of stainless steels in preventing fouling and marine growths that introduce tight crevices. For levels greater than about 0.1 to 0.2 mg L^{-1} free chlorine, the likelihood of crevice corrosion initiation increases [22]. The likelihood of crevice corrosion increases with rising temperatures in chlorinated water with no biofilm.

The highest chlorine level that can be used depends on the material and temperature. Norwegian experience suggests an upper temperature limit of 20 °C for super-duplex and 6% Mo alloys with tight crevices (such as screwed couplings) with 1.5 mg L^{-1} chlorine [16]. Some companies have had better experiences with these alloys and permit use up to 30 °C or more for less severe crevices (e.g. flanged joints). The NORSOK [16] limit is considered over-conservative by some, and a combination of service experience and laboratory data indicates that maximum temperature is a function of chlorine concentration, as shown in Table 3.4 [23].

In heat exchangers with seawater on the tube side and no crevices because the tubes are seal welded into the tube plate, the maximum operating temperature is restricted to about 60°C for superduplex and 6% Mo alloys even though the temperature at which pitting would be expected to occur is higher (~70 °C). This is because of the formation of calcareous deposits that start to appear at this temperature, and a 'crevice corrosion' type of corrosion that can occur beneath them [24]. For higher-temperature service, calcareous scale deposition may be controlled through the use of complexing agents such as polymaleic acids or phosphonates. Sulphuric acid treatment is also frequently used. Such treatments must comply with pollutant discharge regulations and may be difficult to apply to once-through systems such as heat exchangers.

Stainless steels become more resistant to crevice corrosion if they are exposed to a less corrosive environment for some time before contact with a more aggressive environment [23,25]. Thus, it is better to increase the chlorine value from zero to the nominal value over a couple of days, or start with intermittent chlorination. Alternatively, temperature can be used to aid film formation on piping located after a heat exchanger by using a regime such as the following [23].

1. Cold, natural seawater for 2 days minimum.
2. Cold, chlorinated seawater for 5 days minimum.
3. Hot, chlorinated seawater thereafter.

It is well recognised that the performance of welds can be improved by nitric/hydrofluoric (HF) acid pickling, either by immersion, or using a pickling paste or gel [25,26].

Table 3.4 Suggested temperature limits for superduplex stainless steel in seawater at different chlorine concentrations [23]

Chlorine concentration (mg L^{-1})	Maximum temperature (°C)
0.7	40
1.0	30
5.0	20
200	10

Stress corrosion cracking

Chloride SCC is not a significant problem in seawater. If it occurs, it is usually in areas of high applied or residual stress such as expansion joints, bolting, or circumferential welds and/or areas where chlorides concentrate by evaporation on a hot metal surface. This includes conditions found in hot climates. Chloride SCC is essentially transgranular, frequently propagates from pits, and is unlikely to occur at temperatures below about 60 °C with austenitic stainless steels such as UNS S31603 (316L). More highly alloyed austenitic and duplex stainless steels do not usually develop chloride SCC below 100 °C.

In austenitic stainless steels, SCC is primarily related to nickel content. Type 300 stainless steels are the most susceptible grades. Alloys with 25% nickel are significantly more resistant than UNS S31603 (316L).

Duplex stainless steels, with mixed ferritic and austenitic structures, have a far greater resistance to SCC than the austenitic grades of similar pitting/crevice corrosion resistance. Ferritic grades with little or no nickel also have very high resistance to chloride SCC. The susceptibility of precipitation-hardened stainless steels to chloride SCC is a function of temperature and the strength level to which they have been aged [27].

Galvanic corrosion

Passive stainless steels are toward the more noble end of the galvanic series and are more noble than copper alloys, aluminium, and steel. Carbon steel and aluminium provide good protection to UNS S31603 (316L), but copper alloys do not provide galvanic protection [18]. The potential of stainless steels, high-nickel alloys (such as UNS N06625 [625] and UNS N10276 [C-276]), and titanium is similar, but graphite can be more noble and cause galvanic corrosion of stainless steels [21,28]. Galvanic corrosion tends to be lower in chlorinated vs. natural seawater [18]. More information on galvanic corrosion is discussed in Chapter 8.

The high density of inclusions in UNS S30300 (303) and UNS S30323 (303Se), free machining grades of stainless steel, creates numerous galvanic cells in the material; these grades should not be used in seawater [12]. They can fail within 6 months even in contact with aluminium or steel. Graphite-containing gaskets, packing, and lubricants are all responsible for serious galvanic corrosion of stainless steel in seawater and should not be used [18]. Information on the problems with gaskets and stainless steels, along with recommendations on suitable gaskets to minimise corrosion, is discussed in an article by Francis and Byrne [28]. If localised corrosion initiates in a stainless steel, the alloys become more active and, if the contact metal is nobler, the local corrosion rate may increase further.

Microbiologically influenced corrosion (MIC)

When initially immersed in seawater, most stainless alloys exhibit a corrosion potential of approximately 0 mV SCE. As microbes colonise the surface, its potential is ennobled, eventually rising to +250 or even +350 mV SCE. This electropositive potential may lead to crevice corrosion of lower-alloyed stainless steels. Although MIC is occasionally identified in lower-alloy stainless steels and particularly under extended stagnant conditions, it is not identified in practice in the more corrosion-resistant alloys (CRAs) such as 6% Mo austenitic or 25% Cr superduplex alloys [29,30]. The 6% Mo austenitic alloys are used in the United States in power stations to handle brackish water that has caused MIC in lower-alloy stainless steels. The U.S. Navy has also used 6% Mo alloys to prevent MIC in on-board cooling water pipes that alternate between flowing and stagnant conditions. Superduplex stainless steel piping was used with great success on an oil tanker in Australia to combat MIC.

Cathodic protection and hydrogen embrittlement

There are instances when stainless steels are present in systems for components that are cathodically protected. CP can be applied to less noble stainless alloys that do not have sufficient corrosion resistance in seawater, an example being to protect the seawater side of alloy UNS S32205 (2205) subsea flow lines. Also, stainless steels can be present in a predominantly lower-alloy system in which the less noble alloys require CP.

When CP is applied, the rate of hydrogen production is negligible at potentials more positive than ~–800 mV SCE. At more negative potentials, hydrogen is generated on the stainless steel surface. This hydrogen may enter the metal and cause hydrogen embrittlement (HE) of some alloys. The likelihood is greater if the alloy is cold worked and under high mechanical stress. Austenitic stainless steel is better than duplex in resisting HE. Care has to be exercised in the design of CP systems and achieving optimum microstructures when duplex stainless steels are used. The Engineering Equipment and Materials Users' Association (EEMUA) guidelines describe the problem and provide advice on how to minimise the likelihood of HE of duplex alloys under CP [31]. Superferritic stainless steels are especially sensitive to HE caused by CP.

Good fabrication and installation practices

In addition to the factors already discussed, optimum performance from stainless steels, as with other materials, relies on control of the welding parameters [3–5], and attention to detailed design, plus good handling and fabrication practices [32]. The main objective of these welding parameters is to avoid crevices that could cause problems during service. The more important include:

- thorough degreasing is required before welding;
- marks from oil, crayons, sealant, sticky deposits (including stick-on labels), slag, arc strikes, and weld spatter should be removed;
- tooling, blasting, and grinding operations that can leave embedded iron or steel should be avoided; or, if unavoidable, a final nitric/HF pickling treatment should be used;
- correct selection of over-alloyed consumables for welding duplex and super austenitic; and
- inert gas back-purging is necessary during welding to minimise heat tint. Where formed, surface oxides and heat tint should be removed.

For tubular heat exchangers, crevice corrosion can be avoided if the seawater is on the tube side and not the shell side, and the tube is welded into a galvanically compatible tube sheet [33].

QA/QC

It is important to realise, when specifying high-alloy stainless steels, such as 6% Mo and superduplex, that it is not sufficient to just order them to the relevant ASTM standard. This contains only the composition, the minimum mechanical properties, and some basic heat treatment requirements. It has been pointed out that more is required to obtain these materials in a quality that provides ease of fabrication and successful use in service [34,35].

There are, unfortunately, far too many instances of significant problems caused by purchasing material of inferior quality [36]. Byrne et al. suggest some basic QA specifications to ensure that superduplex stainless steel performs as expected [34]. It is important to define similar specifications for other high-alloy stainless steels.

Acknowledgements

The author would like to thank Carol Powell and Ulf Kivisakk for their contributions in the writing of this chapter.

References

1. R. M. Kain et al.: Stainless Steel World 2000 Conference & Expo, Zutphen, Netherlands, 2000, KCI Publishing B.V., p. 28.
2. T. G. Gooch: *Welding J.* 1996, **75**, (5) Suppl, 135.
3. 'Welding guidelines Zeron 100 UNS S32760', Rolled Alloys Bulletin No. 105, Rolled Alloys, Temperance, MI, 2009.
4. C.-O. Pettersson and S.-A. Fager: 'Welding practice for the Sandvik stainless steels SAF 2304, SAF 2205, and SAF 2507', Sandvik Steel info S-91-57-ENG, AB Sandvik Steel, Sandviken, Sweden, 1995.
5. 'How to weld Type 254 SMO stainless steel', Outokumpu Stainless Inc., Schaumburg, IL, 2004.
6. R. Davidsson, J. Redmond and B. Van Deelen: 'Welding of 254 SMO austenitic stainless steels, weldability of materials', Outokumpu Stainless Inc., Schaumburg, IL, 1990.
7. 'Stainless steels: Tables of Technical Properties', Euro Inox, 2nd edn. 2007, www.euro-inox.org
8. G. Schiroky, A. Dan, A. Okeremi and C. Speed: *World Oil*, 2009, **230**, (4), 73.
9. A. Kopliku and C. Mendez: CORROSION 2010, San Antonio, TX, 14–18 March 2010, NACE International, Houston, TX, Paper 10305.
10. H. E. Deverell and I. A. Franson: *Mater. Perform.*, 1989, **28**, (9), 52.
11. N. G. Needham et al.: in 'Stainless steels', 215; 1987, York, UK, The Institute of Materials, Minerals, and Mining [IOM³].
12. 'Guidelines for selection of nickel stainless steels for marine environments, natural waters and brines', Reference Book 11003, Nickel Development Institute, Toronto, Ontario, Canada, 1987.
13. A. Okeremi and M. J. J. Simon-Thomas: CORROSION 2008, New Orleans, LA, 16–20 March 2008, NACE International, Houston, TX, Paper 08254.
14. 'Petroleum, petrochemical and natural gas industries – Materials selection and corrosion control for oil and gas production systems', ISO 21457 (latest revision), Geneva, Switzerland, International Organization for Standardization [ISO].
15. ASTM: 'Standard test methods for pitting and crevice corrosion resistance of stainless steels and related alloys by use of ferric chloride solution', ASTM G48-11, ASTM, West Conshohocken, PA, 2011.
16. 'Materials selection', NORSOK Standard M-001, Rev. 4, August 2004, NORSOK, Lysaker, Norway.

17. C. W. Kovach and J. D. Redmond: CORROSION/93, New Orleans, LA, April 1993, NACE International, Houston, TX, Paper 267.
18. R. Francis: 'Galvanic corrosion: A practical guide for engineers'; 2000, Houston, TX, NACE International.
19. T. Rogne, J. M. Drugli and T. Solen: CORROSION/98, San Diego, CA, March 1998, NACE International, Houston, TX, Paper 696.
20. G. Alabiso et al.: in 'Marine corrosion of stainless steels – Chlorination and microbial effects', 36; 1993, European Federation of Corrosion (EFC) Publication No. 10, London, UK, Maney Publishing.
21. B. Wallen: in 'Avesta Sheffield corrosion handbook'; 2000, Avesta, Sweden, Avesta Sheffield, p. 33.
22. R. E. Malpas, P. Gallagher and E. B. Shone: Stainless Steel World, 87 Vol. Zutphen, The Netherlands, 1987, KCI Publishing BV, p. 253.
23. R. Francis and G. Byrne: Stainless Steel World, Vol. 16, Zutphen, The Netherlands, 2004, KCI Publishing BV, p. 53.
24. P. A. Olsson and M. B. Newman: 'SAF 2507 for seawater cooled heat exchangers', Sandvik Steel S-51-57-ENG, Sandvik Steel, Sandviken, Sweden, 1998.
25. R. N. Gunn, ed.: 'Duplex stainless steels: Microstructure, properties and applications'; 1997, Abington Hall, Abington, Cambridge, UK, Woodhead Publishing Limited.
26. R. Francis and G. Warburton: CORROSION 2000, Orlando, FL, 26–31 March 2000, NACE International, Houston, TX, Paper 630.
27. J. Sedriks: 'Corrosion of stainless steels', 2nd edn, 312; 1996, New York, NY, John Wiley & Sons.
28. R. Francis and G. Byrne: *Mater. Perform.*, 2007, **46**, (10), 50.
29. C. W. Kovach and J. D. Redmond: Stainless Steels '96: Int. Cong. Stainless Steels, Verein Deutscher Eisenhuttenleute, Dusseldorf, Germany, 3–5 June 1996.
30. R. Francis, G. Byrne and H. S. Campbell: CORROSION/99, San Antonio, TX, 25–30 April 1999, NACE International, Houston, TX, Paper 313.
31. 'Guidelines for materials selection and corrosion control for subsea oil and gas production equipment'; 1999, EEMUA 194, London, UK, Engineering Equipment and Materials Users' Association [EEMUA].
32. 'Guidelines for the welded fabrication nickel containing stainless steels for corrosion resistant services', Nickel Development Institute Reference Book, Series No. 11 007, Toronto, Ontario, Canada, Nickel Development Institute, 1992.
33. M. Holmquist, C.-O. Pettersson and M. Newman: 'Guidelines for tube-to-tubesheet joining during fabrication of heat-exchangers in superduplex stainless steel SAF 2507', Sandvik Steel S-91-59-ENG, Sandvik Steel, Sandviken, Sweden, 1996.
34. G. Byrne et al.: CORROSION/2004, New Orleans, LA, 28 March–1 April 2004, NACE International, Houston, TX, Paper 123.
35. R. Francis: 'The selection of materials for seawater cooling systems — A practical guide for engineers'; 2006, Houston, TX, NACE International.
36. G. J. Collie et al.: *Int. J. Mater. Res.*, 2010, **101**, (1), 772–779.

Suggested further reading

G. A. Sussex: Corrosion and Prevention Conference 2008, Wellington, New Zealand, 16–19 November 2008, The Australasian Corrosion Association Inc (ACA), Paper 016.

4
Copper alloys

Carol Powell* and Peter Webster**

Consultants to Copper Development Association

Copper Development Association, 5 Grovelands Business Centre, Boundary Way, Hemel Hempstead HP2 7TE, UK

carol.powell@btinternet.com*, peter.webster@copperdev.org.uk**

4.1   Introduction

Copper-based alloys are traditionally used as wrought products and castings in marine and naval engineering where good resistance to seawater corrosion is required. Copper itself is a versatile metal that has good resistance to corrosion in marine atmospheres and in seawater up to moderate flow velocities. Its properties, in terms of both corrosion resistance and mechanical strength, can be further improved by alloying.

Alloys found in marine service are grouped into brasses (copper–zinc), bronzes (copper–tin–phosphorus, copper–aluminium, and copper–silicon), gunmetals (copper–tin–zinc), copper–nickels, and copper–beryllium. There is a wide range of mechanical and physical properties, and many have excellent anti-galling properties. For applications in seawater systems, copper–nickel and aluminium bronze tend to be preferred, although other alloys are used in marine service and have their specific merits. Copper alloys differ from other alloy groups because copper has an inherent high resistance to biofouling, particularly macrofouling, which can eliminate the need for antifouling coatings or water treatment.

4.2   Application of copper alloys

Typical marine applications using castings and wrought products for each alloy group are included in Table 4.1. The range is wide and includes seawater piping, heat exchangers, pump and valve components, fasteners, bearings, propellers and shafts. The compositions of a selection of alloys from each group are shown in Table 4.2. Associated typical mechanical properties are included in Table 4.3. Higher values can be achieved by cold working or by thermal treatment called age hardening, depending on the particular alloy, product form, and section size.

Copper and Copper Alloys-Standards, Specifications and Applications [1] provides a detailed range of alloy compositions and mechanical properties. Because of the variety of applications for copper alloys, it is important that designers consult with suppliers to clarify what property values and combinations are available to best fit the purpose of the desired product form. The wrought coppers, brasses and bronzes are low-strength in the soft annealed condition, but can be work hardened well into the medium-strength range.

Table 4.1 Typical application of marine copper alloys

Alloy group	Alloy type	Applications
Copper	Phosphorus deoxidised, high-residual phosphorus (DHP)	Copper tubing, nails
Brass	Aluminium brass	Seawater tube and pipe
	Naval brass, Muntz metal	Tube sheet
	Al-Ni-Si brass	Hydraulic, pneumatic, and instrument lines
	Dezincification-resistant (DZR) brass	Through-hull fittings
	Mn bronze (cast and wrought)	Shackles and cabin fittings, propellers, shafts, deck fittings, and yacht winches
Bronze	Phosphor bronze	Springs, bearings, gears, fasteners, rods, and slides
	Silicon bronze	Fasteners (screws, nuts, bolts, washers, pins, lag bolts) staples, and cages
	Aluminium bronze	Sea cocks, pumps, valves, and bushes
	Nickel aluminium bronze (cast and wrought)	Propellers and shafts, pumps, and valves, bushing and bearings, fasteners, and tube plate for titanium tubing in condensers
Gunmetal	Cu–Sn–Zn castings	Pumps and valves, stern tubes, deck fittings, gears and bearings, bollards and fairleads
Copper–nickel	90/10 and 70/30	Heat exchanger and condenser tubes, piping, platform leg and riser sheathing. Seawater cooling and firewater systems. MSF desalination and boat hulls.
	Cu–Ni–Cr	Seawater cast pump and valve components. Wrought condenser tubing
High-strength copper–nickel	Cu–Ni–Al	Shafts, drive bearings and bushes, stab plate connectors and bolting, pump and valve trim, gears, and fasteners
	Cu–Ni–Sn	Bearings, drill components, subsea connectors, valve actuator stems and lifting nuts, subsea manifold and remote-operated vehicle (ROV) lock-on devices, and seawater pump components
Copper–beryllium	Cu–Be	Springs, drill components, subsea cable repeater housings, hydrophones and geophones, subsea valve gates, balls, seats, actuators, lifting nuts, blowout preventer (BOP) locking rings

Age-hardenable alloys such as copper–nickel–aluminium can reach strengths equivalent to ASTM A193 [2] Grade B7 bolting steel. Copper–nickel–tin alloys can reach high strengths too, but the highest strength of any copper alloy is provided by the copper–beryllium alloys that may be hardened by a combination of cold working and age hardening to values comparable to any high-strength steel.

Table 4.2 Nominal compositions of a selection of copper alloys

Alloy group	EN No. or Defence Standard No. (Def Stan)	UNS No.	Cu	Ni	Fe	Mn	Zn	Al	Sn	Other
Copper-DHP	CW024A	C12200	99.9							0.02 P
Al brass	CW702R	C68700	78				Rem	2		0.04 As
Naval brass	CW712R	C46400	61				Rem		1	0.2 Pb
Al–Ni–Si–brass	CW700R	C69100	83	1.2			Rem	1		1 Si
DZR brass	CW602N	C35330	62				Rem			2.0 Pb 0.06 As
Mn Bronze (high-tensile brass)	CW721R	C67500	58		0.8	1.5	Rem	1	0.7	1 Pb
Phosphor bronze	CW453K	C52100	Rem						8	0.3 P
	CW451K	C51000	Rem						5	0.2 P
Silicon bronze	CW116C	C65500	Rem			1				3 Si
Aluminium silicon bronze	CW302G	C64200	Rem					7.4		2 Si
Cast nickel aluminium bronze (NAB)	CC333G	C95800	Rem	5	4.5			9.5		
	Def Stan 02-747 [3]	–	Rem	5	4.5	1		9.2		Ni> Fe
Wrought NAB	Def Stan 02-833 [4]	–	Rem	4.7	4.2	0.3		9.5		
	–	C63200	Rem	4.5	3.8	1.3		9.3		
Gunmetal	CC491K	C83600	Rem	1			5		5	5 Pb
Cu–Ni	CW352H	C70600	Rem	10	1.5	0.7				
	CW353H	C71640	Rem	30	2	2				
	CW354H	C71500	Rem	30	0.7	0.7				
Cu–Ni–Cr	Def Stan 02-824 [5]	–	Rem	30	0.8	0.8				1.8 Cr
		C72200	Rem	16	0.7	0.7				0.5 Cr
Cu–Ni–Al	High-strength Cu–Ni	–	Rem	14.5	1.5	0.3		3		
	Def Stan 02-835 [6]	C72420	Rem	15	1	5		1.5		0.4 Cr
Cu–Ni–Sn	–	C72900	Rem	15					8	
Cu–Be	CW101C	C17200	Rem							1.9 Be

4.3   Alloy groups – Metallurgy and main characteristics

4.3.1   Coppers

Coppers essentially contain more than 99.9% copper and have been used in marine environments for centuries. They have a high resistance to macrofouling but can be subject to erosion–corrosion when the seawater velocity exceeds certain limits. The typical mechanical property range is shown in Table 4.3.

Table 4.3 Typical mechanical properties of some copper alloys commonly used in seawater

Alloy	EN No. or Defence Standard No.	UNS No.	0.2% Proof strength, N mm^{-2}	Tensile strength, N mm^{-2}	Elongation, %	Hardness, HV (unless indicated otherwise)
Copper-DHP (half hard)	CW024A[a]	C12200	180	240	20	70
Al brass	CW702R	C68700	140	350	30	85
Naval brass	CW712R	C46400	140	370	35	105
Al–Ni–Si brass	CW700R[b]	C69100	223	430	45	130
DZR brass	CW602N	C35330	150	350	20	90
High-tensile brass (Mn bronze)	CW721R	C67500	250 min[c]	50 min[c]	14 min[c]	140
Phosphor bronze	CW453K	C52100	280	450	26	130
Silicon bronze	CW116C	C65500	260	385	40	170
Cast NAB	CC333G	C95800	280 min[c]	650 min[c]	12 min[c]	150 HB
	Def Stan 02-747 [3]	–	250 min[c]	620 min[c]	15 min[c]	
Wrought NAB	Def Stan 02-833 [4]	–	245 min[c]	620 min[c]	15 min[c]	
		C63200	275 min[c]	620 min[c]	15 min[c]	
Gunmetal	CC491K	C83600	110 min[c]	230 min[c]	10 min[c]	65 HB min[c]
Cu–Ni	CW352H	C70600	140	320	40	85
	CW353H[b]	C71640	175	450	35	110
	CW354H	C71500	170	420	42	105
Cu–Ni–Cr	Def Stan 02-824 (sand cast) [5]	–	300 min[c]	480 min[c]	18 min[c]	85
		C72200[b] (wrought)	110 min[c]	310 min[c]	46	
Cu–Ni–Al	High-strength Cu–Ni	–	555 min[c]	770 min[c]	12 min[c]	229 min[c]
	Def Stan 02-835 [6]	C72420	430 min[c]	725 min[c]	18 min[c]	170
Cu–Ni–Sn (spinodally hardened)	–	C72900	1035	1137	6	326
Cu–Be	CW101C	C17200	700	850	12	222
	Aged		900 min[c]	1100 min[c]	3 min[c]	354

[a]Half hard.
[b]Tube only.
[c]Minima where stated can vary with product form and section thickness.

Strength and hardness are increased from the annealed condition by cold work. They have excellent thermal and electrical conductivity and good corrosion resistance in marine atmospheres and in seawater, corroding evenly and showing little pitting and crevice corrosion.

4.3.2 Brass alloys

These copper–zinc alloys usually have small additions of other elements to enhance their properties, for example, arsenic or tin for inhibition of dezincification, or lead to aid pressure tightness and/or machining operations [7]. They are divided into two groups for seawater service:

1. Single-phase (alpha) brasses that have up to 37% Zn.
2. Two-phase (alpha beta) brasses that start to form above about 37.5% Zn.

Two main corrosion issues that should be taken into account when selecting and designing with brass alloys are dezincification and ammonia stress corrosion cracking (SCC). Additions of arsenic can successfully inhibit dezincification in alpha brasses, whereas tin can slow down the process in both groups of brass alloys. SCC will be discussed in Section 4.4.3.5

Alpha brass alloys are tough, more ductile than copper, and can be readily cold worked. Their strength increases with zinc levels and cold work. Aluminium brass is one of the more well-known alloys in this group and is sufficiently resistant to erosion–corrosion to be used for piping and heat exchangers/condensers. Alloys with up to about 15% Zn (with the precise content depending on other elements present) are immune to dezincification. Higher-zinc alpha brasses can be successfully used in clean seawater as long as they are inhibited against dezincification by 0.02–0.06% arsenic.

Additionally, UNS C69100 (CW700R) is an alpha brass with about 1% each of aluminium, nickel, and silicon. It has been successfully used in marine environments, particularly for hydraulic control and instrumentation lines up to 35 MPa. It can be precipitation hardened if required, performs well in seawater and marine atmospheres, and has high resistance to dezincification.

Alpha beta brasses are alloys that are hot worked. Other elements can be added, such as lead to improve machineability, and additions of Al, Sn, and Mn produce a range of high-tensile-strength brasses that are readily hot rolled, forged, extruded, and cast. Alpha beta brasses are subject to dezincification [7] in seawater (e.g. Muntz metal; UNS C28000 [CW 509L]), although tin additions (e.g. naval brass: UNS C46400 [CW 712R]) can slow this down considerably.

Cast and wrought manganese bronze alloys also fall into this group; the name is a misnomer as they are essentially high-tensile brasses. Alloys with about 3% Mn and similar amounts of aluminium and nickel provide good service as medium-duty propellers; however, cathodic protection (CP) is required to avoid dezincification.

Dezincification-resistant (DZR) brass (UNS C35330 [CW602N]) was developed to provide a brass that is two-phase, and therefore good for hot working, but can be converted to an all-alpha structure by heat treatment. The presence of arsenic makes the alloy resistant to dezincification by stabilising the alpha phase. Although developed for domestic plumbing service, it can be used in seawater, too, and is approved by Lloyds Register for through-hull fittings [7,8] on yachts and small craft.

4.3.3   Bronze alloys

The term *bronze* originally applied to Cu–Sn alloys. This is not as appropriate now because the addition of small quantities of other elements is found to increase its strength as in Cu–Sn–Zn alloys (gunmetals) and Cu–Sn–P (phosphor bronzes).

The search for higher strength alloys also led to alloys that do not contain tin but have adopted the concept of a superior alloy by including the word 'bronze,' for example, silicon bronze and aluminium bronze. Bronze alloys have high resistance to ammonia SCC compared with brass alloys [8].

4.3.3.1   Phosphor bronze

Binary copper–tin alloys can be rolled and drawn to increase their strength and hardness by cold work. The mechanical properties can be further increased by small additions of phosphorus (up to 0.4% may be present). Castings may contain more than 8% Sn and if so, may require soaking at temperatures of about 700 °C until a second tin-rich phase disappears, returning it to a more corrosion-resistant single-phase alloy. Phosphor bronze alloys corrode evenly and have little tendency to pit. In general, the higher the tin content, the higher the resistance to seawater corrosion. The higher tin-bronzes have good resistance to seawater polluted with sulphides when compared to other copper alloys [8].

4.3.3.2   Gunmetals

Gunmetals are tin bronze castings containing 2-10% zinc and may be further modified by addition of lead (up to 8%) and nickel (up to 6%). The name gunmetal derives from their use, at one time, for gun barrels. They have been traditionally used for marine components such as centrifugal pump impellers, valve seats, taps and pipe fittings. They are not prone to dezincification, SCC, or pitting, nor is crevice corrosion a problem. For use in seawater, sound casting practices and low levels of porosity are necessary [9]. It is preferable to choose a gunmetal with a tin content above 5% and with a low percentage of lead; the more common grades have 5, 7, or 10% tin.

Lead contents of up to 6% have little effect on the corrosion resistance of gunmetal under atmospheric conditions and in normal fresh and seawater at moderate flow rates. When the flow velocity is high, less than 3% lead in such components as centrifugal pump impellers may be advantageous [9]. The addition of lead ensures pressure tightness so they can be used for valve bodies and pump casings. At levels of about 10% lead, so called leaded tin-bronzes make excellent bearing alloys for use in rotating components.

4.3.3.3   Silicon bronze

The most common silicon bronze contains about 3% silicon and 1% manganese and has very good seawater corrosion resistance, and resistance to SCC by ammonia. Silicon bronze has a long history of use as fasteners (screws, nuts, bolts, washers, pins, lag bolts, and staples) in marine environments, including screws used in wooden sailing vessels.

Silicon bronzes have an alpha phase metallurgical structure. They generally have the same corrosion resistance as copper but with higher mechanical properties and superior weldability. The silicon provides solid solution strengthening. They are tough, with high resistance to shock and galling.

4.3.3.4   Aluminium bronze

These alloys are basically copper with 4–12% Al and have a thin, adherent surface film of copper and aluminium oxides that heals very rapidly if damaged. They have good resistance to corrosion, erosion, and wear, as well as good mechanical and corrosion fatigue properties.

At less than 8% Al, the alloys are alpha phase and can be readily rolled and drawn. At 8–12.5% Al, a second phase, beta, is formed, and the alloys can be wrought or cast. Additions of iron, manganese, nickel, or silicon can also be present. Generally, the corrosion resistance of the aluminium bronzes increases as the aluminium, and other alloying additions, increase. Thus, nickel–aluminium–bronze (NAB) in both wrought and cast forms is highly corrosion-resistant in unpolluted waters. A derivation, aluminium silicon bronze (UNS C64200 [CW302G]) (also covered by Defence Standard 02-834 [10]), finds application notably when low magnetic permeability is required.

The metallurgical structure of alpha plus beta aluminium bronze alloys is very complex and described in detail together with corrosion properties and applications by Meigh [11] and Campbell [12]. Careful control of chemistry and processing is required to ensure that the structure is maintained in an optimum condition and does not form less corrosion-resistant phases that can promote selective phase attack.

Nickel and iron are two alloying additions which assist this by modifying the transformation of beta phase to less harmful phases to provide and maintain high corrosion resistance. They also increase ductility and improve castability. These alloys are known as the nickel aluminium bronzes (NABs). Those with about 5% each of nickel and iron are widely used for their strength and corrosion resistance and exhibit useful precipitation-hardening characteristics.

Even so, if the NAB cools too quickly, such as occurs during welding or with thinner sections, not all of the beta phase may transform. This became of particular relevance to naval applications and final heat treatments were developed to counteract this; NAB sand castings to Defence Standard 02-747 Part 2 are required to be heat treated for 6 h at 675 ± 15 °C and air cooled [3]. Heat treatment is also applied to wrought products; for example, NAB extruded rod and bar stock up to 40 mm, and rolled rod and bar up to 30 mm, in accordance with Defence Standard 02-833 Part 2, require a heat treatment at 740 ± 20 °C followed by an air cool [4].

4.3.4   Copper–nickel alloys

4.3.4.1   90/10 and 70/30 copper–nickel alloys

Of the wrought copper alloys, the copper–nickel alloys, and in particular the more economical 90/10 alloy, are the most widely used for seawater systems (piping and heat exchangers/condensers), and are also used for boat hulls and splash zone sheathing on offshore structures [13]. The 70/30 alloy is stronger and can withstand higher flow rates.

General corrosion rates in seawater are normally 0.002 to 0.02 mm year^{-1}, decreasing to the lower end of the range with time. They have high resistance to chloride pitting, crevice corrosion, and SCC and do not have localised corrosion limitations caused by temperature. Piping typically is used up to 100 °C. Copper–nickels have a high resistance to biofouling, particularly the 90/10 alloy [14].

There is also a modified 30% Ni alloy containing 2% Mn and 2% Fe (UNS C71640 [CW353H]) that is only commercially available as condenser tubing. It was developed for high resistance to erosion–corrosion, particularly in the presence of suspended solids. It is extremely successful in multistage flash desalination plants, notably in the heat rejection and brine heater sections [13].

Ammonia SCC in seawater or sulphide stress cracking/hydrogen embrittlement (HE) are not problem areas with these copper–nickels. However, care is still required in polluted conditions as ammonia can lead to higher corrosion rates and can also cause low-temperature hot-spot corrosion in heat exchanger tubes [8]. Sulphides can cause pitting and higher corrosion rates, usually in situations when aerated water mixes with sulphide-containing waters. An established oxide film offers a good degree of resistance to such corrosion, as does ferrous sulphate dosing [8].

These alloys are ductile and can be welded. If weld consumables are used, the 70/30 Cu–Ni electrodes and filler metals are normally preferred [15] for both 90/10 and 70/30 alloys. No post-weld heat treatment is required to maintain corrosion resistance. Copper–nickel can also be welded to steel using the appropriate consumables.

Detailed information about the corrosion performance, mechanical properties, fabrication, and biofouling properties of 90/10 and 70/30 copper–nickel can be found in the *Information sources* section at the end of this chapter.

Further, copper-nickel alloys containing chromium have also been developed; a wrought alloy tubing, C72200, to provide higher resistance to erosion corrosion and a cast alloy, Def Stan 02-834[5], which is used by the UK Royal Navy for pumps and valves.

4.3.4.2   High strength copper–nickel alloys

Although copper–nickel alloys have a long history of use in marine environments because of their excellent corrosion properties and good antifouling properties, they have moderate mechanical properties that are improved by cold working. As a means of improving mechanical strength, two principal alloying routes have been followed: the Cu–Ni–Al system in which precipitation hardening allows high strengths and the Cu–Ni–Sn system that relies on spinodal decomposition of the structure. Both types of alloy can achieve high strengths matching that of bolting steel.

4.3.4.2.1   Cu–Ni–Al

In Cu–Ni–Al alloys, the aluminium increases the strength by a conventional precipitation-hardening mechanism, principally consisting of Ni_3Al, otherwise known as gamma prime. Additional elements, such as Fe, Nb, and Mn, are introduced to the basic Cu–Ni–Al ternary alloy to increase the effectiveness of this phase. The 0.2% proof strength levels of about 700 N mm^{-2} are achievable together with good antigalling properties, while retaining low corrosion rates and resistance to HE. The alloys were refined over the years to improve resistance to ammonia SCC [16].

4.3.4.2.2   Cu–Ni–Sn

Cu–Ni–Sn alloys display spinodal strengthening through the development of submicroscopic chemical composition fluctuations. A significant increase in the strength over the base metal results from the spinodal decomposition, with proof strengths typically 690 to more than 1000 N mm^{-2} [17].

The alloys are used subsea where bearing performance, non-magnetic, low-fouling, antigalling, high-strength properties are required, such as for stems, bushes and bearings. Applications with sliding movement and/or good resistance to corrosion and biofouling are favoured. They are weldable with a post-weld heat treatment being required for weldments if strength is a critical requirement. The alloys retain 90% of room-temperature strength at elevated temperatures as high as 300 °C.

Of the copper-based materials available, UNS C72900 is one of the highest strength, low-friction, non-magnetic, non-galling metals that work in most sour service conditions. Its seawater corrosion rate is very low. Resistance to erosion–corrosion in sand-laden seawater is also very good.

4.3.5   Copper beryllium

In its age-hardened condition, copper beryllium attains the highest strength and hardness of any commercial copper-based alloy, while retaining low corrosion rates in seawater and excellent biofouling resistance [18]. The tensile strength can exceed 1300 N mm^{-2} depending on temper, while the hardness approaches HRC 45. It has high galling resistance, and is immune to HE and chloride-induced SCC. Also, in the fully aged condition, the electrical conductivity is a minimum of 22% of the International Annealed Copper Standard (IACS).

4.4   Corrosion characteristics

4.4.1   Marine atmospheres

Copper and copper alloys have a high degree of resistance to atmospheric corrosion. The resistance is a result of the development of a protective surface layer of corrosion products that reduces the rate of attack. When exposed to the weather, copper eventually develops a uniform green patina. Data from exposure tests [19] for up to 20 years in various marine sites examining a variety of copper alloys has found corrosion rates in the range of 1.3 to 26 × 10^{-4} mm year^{-1}. However, the corrosion rate was much higher for alloys susceptible to dezincification [8]. Brass alloys can be susceptible to SCC when atmospheres are contaminated by ammonia, or closely related substances such as amines.

For galvanic couples in the atmosphere, copper is usually the cathode and can accelerate corrosion of steel, aluminium, and zinc. It is usually satisfactory when coupled with stainless steel [8].

4.4.2   Splash zone

Metal sheathing has proved a very successful approach for preventing splash zone corrosion of steel structures and is applied to the region through the splash/spray zone to a short distance below the tidal zone. Two alloys predominate: namely 65/35 nickel–copper (UNS N04400) and 90/10 Cu–Ni. The former has been applied to the legs and risers of offshore structures for the past 50 years, and the 90/10 Cu–Ni was first used about 25 years ago in the Morecambe Gas Field, UK [20]. Corrosion rates for both have been minimal.

The sheathing is normally welded into position as both alloys can be successfully welded directly to steel with appropriate consumables. Galvanic effects on the steel/sheathing junctions are avoided on the lower end by CP on the steel structure and by coatings at the top of the sheathing in the atmospheric zone. When insulated from the steel (e.g. by neoprene or concrete), the copper–nickel can make full use of its biofouling resistance in the more submerged areas. Even with CP, the fouling is reduced, very loosely attached, and easily removed. Experience from the Morecambe Field, UK, where zinc anodes are applied to the legs, has found that fouling growth only reaches 25 mm. A good source of information and downloadable papers on this subject can be found at www.coppernickel.org.

4.4.3   Submerged conditions

There are important differences in the corrosion characteristics of copper alloys in seawater but their general behaviour can be summarised as a low, general corrosion rate in quiet seawater with very little tendency to pit. They also have useful resistance to flowing seawater even at moderately high velocities. Care should be taken to avoid high flow rates and turbulence, as well as exposure to extended polluted conditions when ammonia or sulphides are present.

4.4.3.1   Protective surface film

The corrosion resistance of copper and copper-based alloys in seawater is determined by the nature of the naturally occurring and protective corrosion product film that forms on their surface. These films can be multilayered and complex, and their compositions depend on the alloy group.

Protective surface films form by exposure to clean seawater over the first couple of days, but take longer to mature depending on the seawater temperature, and normally can take up to 2 to 3 months in European and North American waters [21]. Once established, the films can withstand intermittent levels of pollutants such as ammonia and sulphides.

The long-term, steady-state corrosion rate for copper and its alloys is on the order of 0.025 mm year^{-1} or less and for copper–nickel alloys is 0.002 mm year^{-1} within those flow velocities that each alloy can tolerate without damage to the protective corrosion product film. Long-term, steady-state corrosion rates provide a more accurate estimate of service life for copper alloys than data from short-term exposures.

4.4.3.2   Effects of flow

The protective surface film that forms on copper alloys is resistant to flow until the shear stress from the flowing water is sufficient to damage it, leading to much higher corrosion rates. This critical shear stress, or breakdown velocity, varies from alloy to alloy and is also determined by the prevailing hydrodynamic conditions. Breakdown velocities in tube and piping are well understood, and various standards define maximum velocity limits for different bore diameters [22–24]. The order of resistance for tube or piping is Cu < Al brass < 90/10 Cu–Ni < 70/30 Cu–Ni < 2Mn–2Fe Cu–Ni

< Cu–Ni–Cr. Of these, however, the 90/10 Cu–Ni (UNS C70600 [CW352H]) is the most commonly used alloy for seawater piping and heat exchangers having resistance that is sufficiently high to provide good service for many applications.

For geometries that have different hydrodynamics to tubing and piping, the breakdown conditions are less well defined. For example, a typical maximum flow rate within a 90/10 copper–nickel pipe at 100 mm diameter or greater is typically 3.5 m s^{-1} (see Table 4.4), but for a copper–nickel boat hull, speeds of 12 to 19 m s^{-1} have been successful without causing erosion–corrosion. Higher corrosion rates also are tolerable for intermittent flow in pipes. For instance, copper alloys are used very successfully in firewater systems in which water velocities are typically 10 m s^{-1}. The reason excessive corrosion does not occur is because the system often is only in use for an hour or so per week and the surface film readily re-stabilises during the intervals.

In relative terms, copper, silicon bronze, and low-zinc brass alloys have the lowest resistance to erosion–corrosion; higher-zinc brasses are better, and copper–nickels and NAB have the greatest tolerance to higher-velocity seawater.

Castings such as gunmetals (copper–tin–zinc) and aluminium bronzes have good resistance to impingement (erosion–corrosion). Both groups are used for components such as pumps and valve bodies. NAB is also an established alloy for propellers as it has excellent resistance to cavitation.

At higher velocities up to 40 m s^{-1}, most copper alloys show a corrosion rate of approximately 1 to 2 mm year^{-1}. At this water velocity, severe impingement attack is experienced by all of these materials. Consequently, turbulence raisers, such as tight radius bends, partially throttled valves, and localised obstructions that can increase local water velocities by up to five times compared to the nominal flow, should be avoided for all copper alloys.

Minimum flow rates of about 1 m s^{-1} are often included in tube and piping design to avoid sediment deposition that can interfere with heat transfer and lead to sulphide corrosion if sulphate reducing bacteria are present.

Table 4.4 Typical guidelines for minimum and maximum flow velocities of copper–nickel alloys in seawater systems, m s^{-1} [24]

Minimum velocity	Any tube, any alloy – 0.9 m s^{-1}	
Maximum velocities	UNS C70600 (90Cu–10Ni)	UNS C71500 (70Cu–30Ni)
Condensers and Heat Exchangers		
Once-through	2.4	3.0
Two-pass	2.0	2.6
Piping		
≤76 mm ID[a] with long radius[b] bends	2.5	2.8
77 to 99 mm ID[a] with long radius[b] bends	3.2	3.5
>100 mm ID[a] with long radius[b] bends	3.5	4.0
Short radius bends	2.0	2.3

[a]Inside diameter.
[b]For a long radius bend, radius ≥ 1.5 × outer diameter (OD).

4.4.3.3   Selective phase corrosion

Copper–zinc alloys with more than about 15% Zn, such as naval brass and manganese bronze, are prone to selective phase corrosion in seawater. This is a form of corrosion in which the alloy is corroded and replaced by a porous deposit of copper. The rate of attack can be severe (e.g. 20 mm year^{-1} in 60/40 brass), and because the copper deposit is porous and brittle, leakage may occur, for example, in condenser tubing and tube plates. This type of corrosion is called dezincification, and a similar effect is found in aluminium bronze above 9% aluminium (dealuminification). More rarely the effect is found in 70/30 copper–nickel and mostly at temperatures above 100 °C [8].

As discussed in section 4.3.2 *Brass alloys*, single-phase alpha brass alloys can be rendered immune to dezincification by the addition of a small amount of arsenic (normally 0.02 to 0.06%) but this is ineffective in two-phase alloys such as 60/40 brass. However, in these alloys, the addition of 1% tin reduces the rate of dezincification considerably, and this addition is made in naval brass alloys.Therefore, when thick sections are used, as might be the case in tube plates with aluminium brass or copper–nickel heat exchanger tubing, acceptable lives can often be obtained from naval brass without any remedial action. Alternatively, dezincification in brass alloys can be avoided by using CP with iron or zinc anodes, or impressed current CP.

The addition of about 5% each of nickel and iron renders the duplex 10% aluminium bronze alloy more resistant to dealuminification, although the heat affected zone of weld in this material is susceptible in the as-welded condition. Specific heat treatments have been developed in NAB to avoid this, as described earlier in this chapter in section 4.3.3.4.

Denickelification is very rare in 90/10 copper–nickel, and is also infrequent in 70/30 copper–nickel. A combination of low flow rates, deposits, high temperatures, and heat transfer in condensers favours the occurrence of this phenomenon causing thermogalvanic effects often referred to as hot-spot attack. The solution is more frequent cleaning to remove deposits that could lead to hot spots, and/or increasing flow rate to avoid deposits. A similar attack can occur at lower temperatures when ammonia is also present; ferrous sulphate dosing assists in its prevention when this is the case [8].

4.4.3.4   Pitting and crevice corrosion

Copper alloys do not generally suffer from pitting in clean, flowing water, although pits can occur in copper alloys from pollution, by sulphides generated in sediment, or deposits by anaerobic bacteria. While copper alloys are not subject to chloride pitting, heat exchanger tubes can pit if there are excessive residual carbon films in the bore from the manufacturing process; however, this is a rare occurrence with modern manufacturing methods [8].

Crevices can form at shielded areas under deposits and tight metal or non-metal contact areas such as under washers, O-rings, and flanged connections. In copper alloys, problems associated with crevice corrosion are rarely observed. This is because the mechanism is a 'metal ion concentration cell' effect with the area within the crevice becoming more noble than that surrounding it. This is very different from that associated with stainless steels as any resulting corrosion in copper alloys is mild and shallow in nature, occurs outside the crevice and is not temperature dependent.

However, aluminium bronze alloys in seawater can experience corrosion within the crevice [12] which is essentially a type of selective phase corrosion, and the degree depends on the phases present. NAB, which has a complex metallurgical structure [11], becomes susceptible when the pH drops within the crevice, turning the normally noble phase (kappa III) to one that is less noble than the surrounding alpha phase. This is insignificant if the surrounding exposed surface forms a protective film in aerated, flowing seawater at the outset, or the alloy is cathodically protected, or has sufficient galvanic protection from adjacent components [8,11].

4.4.3.5   Stress corrosion cracking

Ammonia and mercury are the primary causative reagents for SCC of copper alloys, which are essentially immune to chloride SCC as well as HE in seawater.

Brass alloys are the most susceptible to ammonia SCC [7]. In practice, SCC in brasses is more prevalent in marine atmospheres rather than under submerged conditions where very high levels of ammonia and stress levels are necessary for it to occur [8]. If SCC is a possibilty, a stress relief anneal (280 to 300 °C) may be required. Cathodic protection (CP) can prevent SCC without risk of HE. Copper–nickel, copper–tin, and copper–aluminium alloys have a much greater ammonia SCC resistance [8]. Copper–nickels have the highest resistance, and the 90/10 and 70/30 alloys are virtually immune in seawater.

Although aluminium bronze is used satisfactorily in seawater, the matching weld filler metal can be susceptible to SCC in marine and desalination environments. The remedy is to weld with a higher, 10% Al, duplex filler metal and make a final pass with NAB filler to minimise the potential for galvanic corrosion. NAB propellers are routinely welded with NAB filler without increasing the susceptibility to SCC in service. Manganese bronze propellers can be stress-relief annealed after weld repair to reduce the likelihood of SCC.

4.4.3.6   Galvanic corrosion

Copper alloys are central in the galvanic series of metals and alloys in sea water and generally compatible with each other unless area ratios are unfavourable. However, they are less noble than super stainless steels, titanium, and graphite, and this difference needs to be taken into account during design. Chlorination can make this effect less pronounced. Connection to less noble alloys or under CP can significantly reduce the biofouling resistance. More information on galvanic corrosion is provided in Chapter 8 and on the biofouling aspects in section 4.5.13.

4.4.3.7   Polluted seawater and sulphides

If exposed to polluted water, especially if this is the first service water to come in contact with the alloy surface, any sulphides present can interfere with surface film formation, producing a black film containing cuprous oxide and sulphide. This is not as protective as films formed in clean water, and higher general corrosion rates and pitting can be experienced. The sulphide film can be gradually replaced by an oxide film during subsequent exposure to aerated conditions, although high corrosion rates can be expected in the interim. However, if an established cuprous oxide film is already present, then periodic exposure to polluted water can be tolerated without damage to the film.

Exposure to sulphides by copper alloys should be restricted whenever possible and particularly during the first few months of contact with seawater while the protective oxide film is maturing. Ferrous sulphate dosing is used to provide added protection for aluminium brass and copper–nickel seawater systems. In addressing this issue, the UK Ministry of Defence has recently completed a document reviewing guidance and treatment requirements for the mitigation of corrosion of copper alloys in seawater systems aboard surface ships and submarines during build, refit, maintenance, and operation [24]. Much of this is also relevant to other marine and offshore industries.

NAB is successfully used for pumps and valves in clean seawater, but is susceptible to erosion–corrosion in polluted seawater, and in such applications, a 10% tin–bronze may be preferred. Gunmetal, however, is sometimes used for pump and valve bodies because it is easier to produce as pressure-tight castings.

4.4.3.8   Chlorination

Essentially, copper alloys can withstand chlorination levels normally used for controlling fouling in mixed-metal seawater systems. Prolonged, excessive dosing can lead to corrosion and reduced resistance to erosion–corrosion. Chlorine and ferrous sulphate dosing should be staggered and not occur at the same time because there is a tendency for a precipitated floc to be formed. Low levels of chlorination are used to mitigate attack in sulphide-polluted waters [8].

4.5   Galling and seizing

Copper-based alloys are well known for their wear resistance, but less known for their outstanding resistance to galling and seizing [21,25]. This includes high-strength copper–nickels and copper–beryllium. The tin–bronzes are used for components in the sliding mechanisms of naval ordnance that must resist galling and seizing under the extreme impact loadings of recoil. Lubrication is often marginal or non-existent. NAB and other cast copper alloys are widely used for wear rings. Copper alloys are also used for applications such as valve seats, propeller shafting, and ships' stern tube bearings for which high resistance to galling and seizing in seawater is a principal consideration.

4.6   Marine fouling

Biofouling is commonly found on marine structures, including pilings, offshore platforms, boat hulls, and even within piping and condensers. The fouling is usually most widespread in warm conditions and in low-velocity (<1 m s^{-1}) seawater. Above 1 m s^{-1}, most fouling organisms have difficulty attaching themselves to surfaces. There are various types of fouling materials and organisms, including natural organic matter, bacteria and protozoa, plants (slime algae), hydroids, tunicates, sea anemones, barnacles, and molluscs (oysters and mussels). In steel, polymers, and concrete marine construction, biofouling can be detrimental, resulting in unwanted excess drag on structures and marine craft in seawater, or causing blockages in pipe systems.

Marine organisms attach themselves to some metals and alloys more readily than they do to others. Steels, titanium, and aluminium foul. Many copper-based alloys have a high inherent resistance to biofouling. This is particularly so for macrofouling (grasses and shell fish), although microfouling (slimes) still occurs, often to a reduced extent [13,26]. When exposed for long periods under quiet conditions, some macrofouling can eventually colonise, but this is observed to slough away at intervals and can be readily removed by a light wiping action. The most important requirement for optimum biofouling resistance is that the copper alloy should be freely exposed or electrically insulated from galvanically less noble alloys and CP [13]. The 90/10 copper–nickel alloy has found several applications that take advantage of its corrosion and biofouling resistance, including markers and sheathing on the legs of offshore structures [13,14,20,26]. Special brass alloys are also being used as woven mesh aquaculture cages.

Acknowledgements

Thanks are given to Copper Development Association UK, KME Germany, Columbia Metals UK, Materion Brush Ltd, Bolton Metals and Roger Francis for the information and support provided in writing this chapter.

References

1. Copper Development Association (CDA): 'Copper and copper alloys – Standards, specifications and applications', Publication 120, CDA, Hemel Hempstead, UK, 2004.
2. ASTM: 'Standard specification for alloy-steel and stainless steel bolting for high temperature or high pressure service and other special purpose applications', ASTM A193/A193M ASTM, West Conshohocken, PA, 2011.
3. 'Requirements for nickel aluminium bronze castings and ingots', Defence Standard 02-747, Part 2 and Part 4, MoD, London, UK.
4. 'Requirements for nickel aluminium bronze. forgings, forging stock rods and sections', Defence Standard 02-833, Part 2, MoD, London, UK.
5. 'Copper nickel chromium sand casting and ingots', Defence Standard 02-824, Part 1 MoD, London, UK.
6. 'Requirements for High Strength Copper Nickel Manganese Alloy Forgings, Forging Stock Rods and Sections', Defence Standard 02-835, Part 2, MoD, London, UK.
7. CDA: 'The brasses – Properties & applications', Publication 117, CDA, Hemel Hempstead, UK, 2005.
8. R. Francis: 'Corrosion of copper and its alloys – A practical guide for engineers'; 2010, Houston, TX, NACE International.
9. F. Hudson and D. Hudson: 'Gunmetals castings'; 1967, London, UK, Macdonald and Janes.
10. 'Requirements for aluminium silicon bronze', Defence Standard 02-834 Part 2, MoD, London, UK.
11. H. J. Meigh: 'Cast and wrought aluminium bronzes – Properties, processes and structure'; 2000, London, UK, The Institute of Materials, Minerals, and Mining [IOM³] and Maney Publishing.
12. H. Campbell: 'Aluminium bronze corrosion resistance guide,' Publication 80, CDA, Hemel Hempstead, UK, 1990.
13. C. Powell and H. Michels: CORROSION/2000, Orlando, FL, 26–31 March 2000, NACE International, Houston, TX, Paper 627.
14. W. Schleich and K. Steinkamp: Stainless Steel World 2003 Conference, Maastricht, The Netherlands, November 2003, Paper PO739.
15. 'Copper–nickel fabrication: Handling, welding, properties, resistance to corrosion and biofouling, important applications', joint publication by Copper Development

Association Inc., Publication No. A7020-XX/99, Nickel Institute, Publication No. 12014, and CDA UK, Publication No. 139.

16. C. D. S. Tuck: 'High strength copper nickels', Langley Alloys, CDA, Hemel Hempstead, UK, 2008.

17. R. E. Kushna et al.: 'Engineering guide. Toughmet 3AT (C72900) and 3CX (C96900)', Brush Wellman Inc., Cleveland, OH, 2002.

18. 'Guide to beryllium copper', Brush Wellman Inc., Cleveland, OH, 2007.

19. C. D. S. Tuck, C. A. Powell and J. Nuttall: in 'Shreir's Corrosion', (ed. J.A. Richardson *et al.*), Vol. 3, 1937–1973; 2010, Amsterdam, Elsevier.

20. C. Powell and H. Michels: EUROCORR 2006, Maastricht, The Netherlands, 24–28 September 2006. EFC Dechema, Germany.

21. A. H. Tuthill: 'Guidelines for the use of copper alloys in seawater', Joint Publication by the Copper Development Association, and Nickel Institute. CDA, New York; NI, Toronto, 1987.

22. BSI: 'Salt water piping systems in ships', BS MA 18, London, UK The British Standards Institution.

23. 'Pipelines of copper–nickel alloys – Part 2. Basic principles for design, fabrication and testing', German Standard DIN 85004, Berlin, Germany, German Institute for Standardization [DIN].

24. 'Protection of seawater system pipework and heat exchanger tubes in HM surface ships and submarines', Defence Standard 02-781, Issue 2, MoD, London, UK.

25. CDA: 'Copper alloy bearings. Properties and applications', CDA Publication TN 45, CDA, Hemel Hempstead, UK, 1992.

26. S. Campbell, R. Fletcher and C. Powell: Proc. 12th International Congress on Marine Corrosion and Fouling, Southampton, UK, 27–30 July 2004.

Suggested further reading

CDA: 'Copper and copper alloy castings. Properties and applications', CDA Publication TN 42, CDA, Hemel Hempstead, UK, 1991.

R. Francis: 'The selection of materials for seawater cooling systems – A practical guide for engineers'; 2006, Houston, TX, NACE International.

F. LaQue: 'Marine corrosion: Causes and prevention'. Corrosion monograph series; 1975, Hoboken,NJ, John Wiley and Sons.

T. H. Rogers: 'Marine corrosion'; 1968, London, UK, George Newnes Limited.

N. Warren: 'Metal corrosion in boats: The prevention of metal corrosion in hulls, engines, riggings and fittings', 2nd edn; 1998, Hampshire, UK, Adlard Coles Nautical.

CDA: 'Copper alloys for marine environments' Publication 206, CDA, Hemel Hempstead, UK; 2011.

Information sources

For detailed information about copper–nickels, see downloadable papers on www.coppernickel.org

Technical publications on corrosion, fabrication, and properties of copper alloys are available from the Copper Development Association websites: www.copperinfo.co.uk and www.copper.org

5
Nickel alloys

Carol Powell* and David Jordan**

Consultants to the Nickel Institute
Nickel Institute, The Holloway, Alvechurch, Birmingham, B48 7QA, UK
carol.powell@btinternet.com*, dejordan@compuserve.com**

5.1 Introduction

Nickel-based alloys used in marine environments are normally derived from the Ni–Cr–Mo or Ni–Cu alloying systems. For the purposes of this overview, a nickel alloy is an alloy with more than 40% Ni. The alloy range includes some of the most highly corrosion-resistant alloys (CRAs) available and, because they are relatively expensive, they tend to be used in more critical and severe applications. For thicker-section components, some applications have involved rolled-clad nickel alloy/steel plate, or a weld overlay layer. The weld consumables are also used to provide over-matching welds when joining lower alloys to maintain corrosion resistance and weld stability, or to avoid the effects of iron dilution.

5.2 Characteristics of nickel alloys

High nickel content provides alloys with a tough, ductile crystal structure that can be used down to cryogenic temperatures without fear of embrittlement. Nickel alloys also have a very high resistance to chloride stress corrosion cracking (SCC), with a virtual immunity above 50% nickel. Mo and W are solid-solution strengthening alloy additions that improve the overall strength of the alloys, as well as their resistance to localised corrosion. The alloys can be work hardened but some (e.g. UNS N05500, UNS N07725, UNS N09925, and UNS N07718) purposely have small titanium and aluminium alloying additions that allow them to be age hardened by thermal treatment to further increase their mechanical strength without loss of corrosion resistance. Niobium is sometimes used for the same purpose (UNS N07718 and UNS N07725).

The Ni–Cr–Mo alloys rely on a protective, predominantly chromium oxide, surface film for their corrosion resistance. It is robust and self-repairing in most marine environments. General corrosion rates are negligible, but some of the lower alloys can be susceptible to chloride pitting and crevice corrosion under quiet conditions. However, the higher Mo grades have very high corrosion resistance, and localised corrosion is unlikely to occur in the marine/seawater environment. In more extreme conditions, higher levels of chlorination or temperature and more acidic pH levels can be tolerated than with many other alloys.

The alloys are weldable, which has led to their use as a corrosion-resistant barrier in the form of a weld overlay in critical areas. The weld consumables are also used for welding lower alloys and stainless steels to provide a noble, strong, and corrosion-resistant weld metal. Typical marine applications are included in Table 5.1.

Table 5.1 Typical marine applications for nickel alloys

Alloy group	UNS no.	Common name	Application
Ni–Cu	UNS N04400	Alloy 400	Propeller shafting, fasteners and connectors, and demisters; Pump and valve components; and Splash zone leg and riser sheathing, and on-board piping
	UNS N05500	Alloy K-500	Pump and propeller shafts, valves and controls, and fasteners
Ni–Cr–Mo	UNS N08825	Alloy 825	Expansion bellows, valves, exhaust systems, and flow lines
	UNS N07718	Alloy 718	Fasteners, springs, marine turbines, and pump shafts
	UNS N06625	Alloy 625	Weld overlays, valves, fasteners, springs, seawater systems, pump shafts, and expansion bellows
	UNS N07725	Alloy 725	High-strength fasteners and
	UNS N09925	Alloy 925	pump shafts
	UNS N10276	Alloy C-276	Heat exchangers
	UNS N06059	Alloy 59	Corrosion-resistant overlays, heat
	UNS N06022	Alloy 22	exchangers, high levels of chlorination,
	UNS N06686	Alloy 686	acidic seawater, and high-strength
	UNS N06200	Ni–Cr–Mo–Cu alloy	fasteners

5.3   Types of nickel alloy in marine use

A selection of alloy grades commonly used in marine services and their compositions are provided in Table 5.2. Typical mechanical properties are shown in Table 5.3 where the difference between the age-hardenable and non-age-hardenable types of alloy is apparent.

5.3.1   Nickel–copper alloys

The main nickel–copper alloys used are UNS N04400 (alloy 400), which consists of approximately two-thirds nickel and one-third copper, and UNS N05500 (alloy K-500), which is similar but contains small additions of aluminium and titanium. By carefully heat treating UNS N05500 (alloy K-500), the nickel, aluminium, and titanium combine to precipitate fine particles of $Ni_3(Ti,Al)$ called gamma prime. These form in the alloy matrix, and stiffen it to roughly double the mechanical properties compared with UNS N04400 (alloy 400), while retaining similar corrosion resistance.

Both alloys have low general corrosion rates even at high water flow rates, while being subject to pitting and crevice corrosion under quiet conditions [1,2]. Cathodic protection (CP), or operation in systems of galvanically less noble metals, enhances the performance of these alloys. At seawater flow velocities greater than 1 m s⁻¹, the surface remains passive and good resistance has been measured up to greater than 40 m s⁻¹. The alloys are typically used for fasteners, shafting, and pump and valve components. Cast versions of UNS N04400 (alloy 400) are available for impellers.

Table 5.2 Composition of some nickel alloys commonly used in marine service

UNS no.	Common name	Nominal composition, wt%							PREN[a]
		Ni	Cr	Mo	W	Nb	Cu	Other	
N04400	Alloy 400	65					32	2 Fe	–
N04401	Alloy K-500	65					30	2.7 Al 0.6 Ti	–
N08825	Alloy 825	42	21	3			2	28 Fe, 0.8 Ti	31
N09925	Alloy 925	44	21	3			1.8	28 Fe, 2.1 Ti, 0.3 Al	31
N00718	Alloy 718	54	18	3		5		18.5 Fe 1 Ti, 0.6 Al	28
N00625	Alloy 625	61	21	9		3.6		3 Fe	51
N07725	Alloy 725	57	21	8		3.5		7.5 Fe, 1.5 Ti 0.3 Al	51
N10276	Alloy C-276	57	16	16	3.5			6 Fe 0.35 V	75
N06022	Alloy 22	56	22	13.5	3			0.35 V	72
N06059	Alloy 59	59	23	16	–			0.5 Fe	76
N06686	Alloy 686	58	21	16	3.7	1			80
N06200	Ni–Cr–Mo–Cu alloy	60	23	16			1.6		76

[a]PREN (pitting resistance equivalent number) is calculated using Equation 5.1.

Table 5.3 Typical mechanical properties of nickel alloys

UNS no.	Common name	0.2% Proof stress, MPa	Tensile strength, MPa	Elongation, %
N04400	Alloy 400	276[a]	552[a]	45
N08825	Alloy 825	310[a]	655[a]	45
N06625	Alloy 625	448[a]	862[a]	50
N10276	Alloy C-276	345[a]	758[a]	60
N06022	Alloy 22	379[a]	793[a]	60
N06059	Alloy 59	340[a]	690[a]	50
N06686	Alloy 686	379[a]	758[a]	60
N06200	Alloy Ni–Cr–Mo–Cu	350[a]	750[a]	65
N05500	Alloy K-500	689[b]	1069[b]	25
N07718	Alloy 718	1138[b]	1413[b]	20
N09925	Alloy 925	827[b]	1172[b]	25
N07725	Alloy 725	896[b]	1276[b]	30

[a]Minimum properties for hot-rolled annealed plate.
[b]Age-hardened. The properties for age-hardenable alloys can vary and typical values are shown.

5.3.2 Nickel–iron–chromium–molybdenum alloys

Nickel–iron–chromium–molybdenum alloys also have very low general corrosion rates and are unaffected by flow rates as great as 40 m s^{-1}. They have a very high resistance to chloride SCC, but can be susceptible to crevice corrosion in seawater unless cathodically protected.

UNS N08825 (alloy 825) is a 42% nickel alloy with 21% Cr and 3% Mo that falls into this category. UNS N09925 (alloy 925) is the age-hardened version. UNS N07718 (alloy 718) was developed for aerospace applications, but its composition with Cr, Mo, and Nb has useful strength and corrosion resistance for offshore environments, and it is widely used in the oil and gas industries. It is its age-hardening properties that allow high strengths to be achieved. In corrosion applications, the alloy is given a different heat treatment than the standard aerospace material to maximise corrosion resistance.

5.3.3   Nickel–chromium–molybdenum alloys

UNS N06625 (alloy 625) was also first developed for high-temperature service in aerospace applications, but it was soon discovered that its composition also gave a high resistance to localised corrosion in aqueous chloride environments and it could also be used as a weld overlay. The alloy is strong but, if even higher strength levels are required, UNS N07725 (alloy 725) is the age-hardenable version giving the strength of UNS N07718 (alloy 718) with the improved corrosion resistance of UNS N06625 (alloy 625). Even higher levels of corrosion resistance can be achieved from UNS N10276 (alloy C-276).

Further developments to maximise levels of Cr, Mo, and W, and, therefore, localised corrosion resistance, while at the same time producing a stable alloy, have since led to UNS N06059 (alloy 59), UNS N06022 (alloy 22), UNS N06686 (alloy 686), and UNS N06200 (Ni–Cr–Mo–Cu). There are subtle differences between the alloys, but all are strong, weldable, and have very high corrosion resistance. They can all be used as weld overlays.

5.4   Corrosion resistance

Potential forms of corrosion in nickel alloys used in marine environments are pitting, crevice corrosion, and SCC. The level of resistance depends on the alloy composition and the severity of the environment.

5.4.1   Marine atmosphere

In the marine atmosphere, Ni–Cu alloys generally have negligible corrosion rates, forming a grey/green surface film over prolonged atmospheric exposure.

UNS N08825 (alloy 825) maintains a bright surface. UNS N06625 (alloy 625) and the high-Mo, Ni–Cr–Mo alloys also essentially retain a bright surface and are unaffected by marine atmospheres.

5.4.2   Splash zone

High corrosion rates of steel structures are found in the splash zone as described in Chapter 2. Ni–Cu alloy sheathing is an effective means of avoiding corrosion in this area when applied from the splash/spray zone to below the tidal zone. UNS N04400 (alloy 400) has been used as a splash-zone sheathing on steel platforms for more than 50 years, and for more than 30 years as a cladding on hot risers [3]. Corrosion rates of the sheathing are minimal. The sheet used is normally 3 to 5 mm thick and welded into position, although the alloy has also been used as metallurgically clad steel pipe for hot risers.

Galvanic corrosion of the steel below the sheathing is suppressed by the operating CP systems, and a properly maintained coating is required at the steel/sheathing interface in the atmospheric zone.

5.5   Immersed

5.5.1   Pitting and crevice corrosion

Pitting and crevice corrosion in Ni–Fe–Cr–Mo and Ni–Cr–Mo alloys normally require the presence of chlorides, as for stainless steels. They are also affected by pH, temperature, and tightness of the crevice. Crevice corrosion occurs more readily than pitting and at lower temperatures in the same bulk environment. Higher levels of chromium, molybdenum, and tungsten improve resistance to both types of corrosion.

Nickel maintains the austenitic-type structure so that high levels of molybdenum can be added while still maintaining a stable alloy. Thus, very high levels of resistance to localised corrosion can be achieved.

The pitting resistance equivalent number (PREN) is a means of comparing the relative resistance of alloys to both pitting and crevice corrosion, and for high-nickel alloys, it can be calculated from their compositions by the same formula as for stainless steels given in Chapter 3 (Equation 3.1). However, nitrogen is not in use as an alloying constituent in nickel alloys because its solubility is too low in these alloy systems. Hence, the PREN for nickel alloys is calculated using the simplified equation shown in Equation 5.1

$$PREN = \%Cr + 3.3\,(\%\,Mo + 0.5 \times \%W) \qquad [5.1]$$

The higher the PREN, the greater the resistance to localised corrosion. Table 5.2 includes the PREN values for the nickel alloys already mentioned.

Ni–Fe–Cr–Mo alloys with a PREN < 40 (e.g. UNS N08825 and UNS N07718) can be susceptible to crevice corrosion in ambient-temperature seawater [4]. Galvanic or cathodic protection is required to ensure their resistance to crevice corrosion during seawater immersion. In polluted conditions when hydrogen sulphide is present, UNS N08825 (alloy 825) has a better resistance to pitting than UNS N04400 (alloy 400), which is known to be more susceptible under that condition [5].

Higher-molybdenum alloys UNS N06625 (alloy 625) and UNS N10276 (alloy C-276) are substantially better and immune to corrosion in many severe marine environments. They retain extremely low corrosion rates in stagnant, flowing, and high-velocity seawater. Although subject to marine attachment, the alloys are highly resistant to crevice corrosion caused by such biofouling. Tight, deep crevices (e.g. sleeves) have occasionally initiated corrosion in UNS N06625 (alloy 625), but the higher alloys, such as UNS N06059 (alloy 59), UNS N06022 (alloy 22), UNS N06686 (alloy 686), and UNS N06200 (Ni–Cr–Mo–Cu), are very corrosion-resistant where extremely aggressive conditions are found. These alloys show little tendency to pit and crevice corrode even under many conditions of higher temperatures, chlorination, and low pH [6,7].

Nickel–copper alloys have been used for seawater service since the early 20th century and were initially considered resistant to pitting and crevice corrosion. This is now thought to be because they were usually the noble component in a mixed-metal system that provided them with galvanic protection [5]. When unprotected in quiet seawater, Ni–Cu alloys UNS N04400 (alloy 400) and UNS N05500 (alloy K-500) can be susceptible to localised corrosion. Pitting tends to slow down after a fairly rapid initial attack, rarely exceeding 1.3 mm deep even after several years of exposure [4]. Connections to more noble high-alloy stainless steels, Ni–Cr–Mo alloys, and Ti can accelerate the localised corrosion of nickel–copper alloys. They can also pit in the presence of pollutants, such as hydrogen sulphide formed in stagnant, putrid conditions [2,5]. Low levels of chlorination ($\sim$0.5 mg L^{-1}), which eliminate the surface biofilm, can reduce the susceptibility to pitting. However, high levels of chlorination can cause pitting, as explained in section 5.5.5 *Biofilms and microbiologically influenced corrosion* later in this chapter [5].

5.5.2   Stress corrosion cracking and hydrogen embrittlement

Type 300 series austenitic stainless steels are prone to chloride SCC at temperatures above $\sim$60 °C when under tensile (applied or residual) stress, and tend to fail in a transgranular manner (i.e. the crack grows through the grains in the metallic structure rather than around them). This can occur with minimal thinning, but can quickly and dramatically penetrate through the metal thickness.

UNS N08825 (alloy 825) was developed to provide high resistance to chloride SCC compared to these low-alloy stainless steels, and, although it can fail under severe laboratory conditions, such as 42% magnesium chloride solution boiling at $\sim$154 °C, it is normally considered to have a very high resistance in the marine environment. Higher-nickel alloys of the Ni–Cr–Mo type and Ni–Cu alloys are not susceptible to this type of corrosion in seawater environments [4].

Flexible bellows are used for accommodating expansion and contraction of pipelines and ducts that commonly carry hot fluids, such as steam containing chloride impurities, and are a prime target for SCC. Readily fabricated alloys such as UNS N08825 (alloy 825) and UNS N06625 (alloy 625) are used for such purposes.

Under highly stressed conditions and CP, hydrogen embrittlement (HE) can occur when hydrogen charging occurs, reducing ductility and fracture toughness. UNS N05500 (alloy K-500) bolting has been known to fail from this mechanism in marine service [8]. UNS N07725 (alloy 725) and UNS N09925 (alloy 925) offer greatly improved resistance to HE [9]. More recently, cold-worked UNS N06059 (alloy 59) and UNS N06686 (alloy 686) have also been considered for fasteners to avoid HE.

5.5.3   Erosion–corrosion

Ni–Cu and Ni–Fe–Cr–Mo alloys have very good resistance to erosion–corrosion from fast-flowing seawater. In the presence of solids, the harder and stronger Ni–Cr–Mo alloys normally have better resistance.

5.5.4   Galvanic corrosion

High-nickel alloys are noble and it is usually the alloy they are connected to that is likely to suffer from galvanic corrosion when area ratios are unfavourable.

Coupling Ni–Cu alloys to copper alloys and steels/cast irons is normally acceptable as long as the area ratios are not extreme; the galvanic protection offered to the

Ni–Cu alloys can minimise localised corrosion. However, UNS N04400 (alloy 400) and UNS N05500 (alloy K-500) are less noble than the Ni–Cr–Mo alloys and high-alloy stainless steels, and can develop localised corrosion as a result if connected to them [5,9].

Ni–Fe–Cr–Mo alloys with a PREN < 40 can also develop pitting and crevice corrosion in seawater and this can be exacerbated if the material is coupled to a higher alloyed nickel alloy, or a high-alloy stainless steel in seawater. See Chapter 8 for more information on galvanic corrosion and measures taken to avoid it.

5.5.5   Biofilms and microbiologically influenced corrosion (MIC)

The biofilms that naturally form on stainless steels and nickel alloys in seawater can lower their relative resistance to localised corrosion compared with those with no biofilm. If low levels of chlorine are added to seawater (~0.5 mg L^{-1}), the biofilm cannot form and corrosion resistance is higher. In more susceptible alloys such as UNS N04400 (alloy 400), this can prevent pitting from occurring when it otherwise would. However, with chlorine levels above 1 mg L^{-1}, the seawater becomes aggressive enough to again cause pitting [5,10].

Biofilms can also encourage the corrosion of lower alloy materials in galvanic couples (see section 8.2.2 *Electrode efficiency* in Chapter 8) compared to some chlorinated situations with no biofilm.

Sulphate reducing bacteria (SRB) that operate under anaerobic conditions are of concern because they can lead to MIC in some nickel alloys in seawater. This normally requires prolonged quiet or stagnant conditions for colonisation to occur. UNS N06625 (alloy 625) and the more highly alloyed alloys are unaffected, but UNS N08825 (alloy 825) and UNS N04400 (alloy 400) may be susceptible to pitting in their presence.

5.6   Weld metals and overlaying

While nickel alloys can generally be welded without difficulty, design and selection of suitable filler metals for corrosive environments require careful consideration. It is particularly important that the weld is at least as resistant to corrosion as the alloy being welded to avoid attack being concentrated on the weld joint. The alloys described in preceding sections are welded with filler metals of similar, but not necessarily identical, composition. However, because of the segregation of alloying elements in weld deposits (which are analogous to castings), the corrosion resistance of welds is inferior to that of a wrought alloy of similar composition [11].

In some applications, it may be necessary to use an 'overmatching' filler metal to ensure adequate resistance to a particular environment. For overlaying, there is the additional factor of excessive dilution of the filler metal deposit by the base metal. This is normally avoided by process modification to reduce the specific heat input and by building up the overlay in several layers.

Nickel alloys can be welded by conventional welding processes. Those most frequently used are manual metal arc/shielded metal arc (MMA/SMAW), tungsten inert gas/gas tungsten arc (TIG/GTAW), and metal inert gas/gas metal arc (MIG/GMAW) welding. MIG/GMAW is more appropriate for workshop application. These can also be adapted for overlaying; for example, the widely used hot-wire TIG process offers high deposition rates coupled with low dilution. Other overlaying processes include submerged arc and electroslag cladding, which use special strip consumables.

As with stainless steels, it is important that nickel alloys be welded under clean conditions in which contamination, especially from carbon and low-alloy steels, is avoided. Materials for fabrication should also be free from contaminants such as oils and grease, coatings and marking inks, especially adjacent to the area where the welding takes place.

All slag should be removed before a fabrication is put into service, although it is not usually essential to eliminate heat tint other than mechanically, for example by grinding. Preheat and post-weld heat treatments are not normally required.

Nickel–copper alloys are welded with a filler metal of similar composition, modified by the incorporation of minor elements to combine with atmospheric oxygen and nitrogen that would otherwise cause porosity in the deposit. Welds in UNS N05500 (alloy K-500) are not as strong as the base metal, but a Ni–Cr–Fe alloy filler metal can be used if this is an important requirement.

While matching filler metals are available for UNS N08825 (alloy 825), they are normally welded with UNS N06625 (alloy 625) consumables for improved corrosion resistance. Similarly, UNS N09925 (alloy 925) is welded with the consumables for UNS N07725 (alloy 725) that are also age-hardenable.

UNS N07718 (alloy 718) is welded by a gas-shielded process with a filler wire of matching composition and, because of its slow rate of hardening, can be age-hardened directly without an intermediate stress-relief heat treatment.

UNS N06625 (alloy 625) consumables deposit weld metals in the same composition range as the wrought alloy, thus with an alloy content sufficient to cope with dilution from lower alloyed materials in dissimilar metal welds without substantial loss of corrosion resistance. They can also be used as overmatching weld metals for conventional stainless steels and 6% Mo stainless steels. Both wire and strip (e.g. for submerged-arc strip cladding) are available for overlaying.

Welding consumables of matching composition are specified for UNS N10276 (alloy C-276) and the more advanced alloys UNS N06059 (alloy 59), UNS N06022 (alloy 22), and UNS N06686 (alloy 686), and all can be used for welds in lower alloyed materials. When conditions such as severe crevices, level of chlorination, and temperature are particularly aggressive for UNS N06625 (alloy 625) overlays, these advanced alloys are also particularly effective as weld overlays in countering attack in areas at risk of severe localised corrosion, for example, in threaded or gasketed joints.

Acknowledgements

The authors would like to acknowledge the Nickel Institute, Roger Francis, Haynes International Inc, Special Metals and ThyssenKrupp VDM for the support and information provided in writing this chapter.

References

1. J. Klower, U. Heubner *et al.*: 'Nickel alloys and high alloy special stainless steels', 3rd edn; 2003, Germany, Expert Verlag.
2. R. Francis: 'The corrosion of copper and its alloys: A practical guide for engineers'; 2010, Houston, TX, NACE International.

3. D. T. Peters, H. T. Michels and C. A. Powell: Int. Workshop on Control for Marine Structures and Pipelines, Galveston, TX, February 1999, American Bureau of Shipping, Houston, TX, 2000, 189–220.
4. 'High performance alloys for resistance to aqueous corrosion', Special Metals Publication No. SMC-026, Special Metals Corporation, Huntington, WV, 2000.
5. R. Francis: 'The selection of materials for seawater cooling systems – A practical guide for engineers'; 2006, Houston, TX, NACE International.
6. Haynes International: 'High performance alloys for seawater service', No. H-2102, Haynes International, Kokomo, IN, 2009.
7. T. Cassagne, P. Houlle and D. Zuili: CORROSION 2010, San Antonio, TX, 14–18 March 2010, NACE International, Houston, TX, Paper 10391.
8. D. Efird: *Mater. Perform.*, 1985, **24**, (4), 37.
9. E. L. Hibner and L. E. Shoemaker: CORROSION/2004, New Orleans, LA, 28 March–1 April 2004, NACE International, Houston, TX, Paper 04287.
10. P. Gallagher, A. Nieuwhof and R. M. Tausk: 'Marine corrosion of stainless steels – Chlorination and microbial effects', 73; 1993, European Federation of Corrosion (EFC) Series, Publication No. 10, London, UK, Maney Publishing.
11. S. McCoy, L. Shoemaker and J. Crum: Stainless Steel World 2001, Stainless Steel World Conference, Zutphen, The Netherlands, 2001. KCI Publishing.

Sources of further technical information

Major high-nickel alloy suppliers provide a comprehensive range of technical data and publications about high-nickel alloys, their properties, fabrication, and application. Information is also available from the Nickel Institute website: www.nickelinstitute.org

6

Aluminium alloys

Clive Tuck

Lloyd's Register EMEA, London EC3M 7BS
clive.tuck@lr.org

6.1   Introduction

Aluminium alloys are normally selected for engineering applications as they have advantages in terms of low density, a high strength-to-weight ratio, high thermal conductivity, acceptable weldability, and, depending on the alloy, good corrosion behaviour in seawater. Although aluminium is a reactive metal, its good corrosion resistance is entirely a result of the rapid formation of a thin, stable, and impervious oxide film on the surface. The thickness of this oxide film, as formed in air, varies between 2.5 and 20 nm depending on the time of exposure. Breakdown of this film, for instance, through chemical attack by chloride ions, results in localised attack of the metal substrate, leading to pitting or crevice corrosion.

6.2   Characteristics of aluminium alloys

6.2.1   Nomenclature of aluminium alloys

Pure aluminium metal is of low strength; therefore, it is used for engineering purposes with the addition of various alloying elements. The different types of alloy produced in this way are grouped into non-heat-treatable and heat-treatable alloys, depending on the microstructures that can be developed to give enhanced strength. For instance, the alloying elements magnesium, manganese, and silicon produce alloys that are strengthened using work-hardening techniques.

The alloying elements copper and various combinations of magnesium, zinc, and silicon can produce heat-treatable alloys. These harden after solution heat treatment and quenching through the development of intermetallic particles (or precipitates) in the material, either by naturally ageing at room temperature, or by artificial ageing at elevated temperature.

The common nomenclature system for aluminium alloys is the one devised by the Aluminium Association and uses the first digit of a four-digit number to define the basic alloy type. The list of general aluminium alloy types is provided below:

- 1xxx – Pure aluminium
- 2xxx – Aluminium–copper
- 3xxx – Aluminium–manganese
- 4xxx – Aluminium–silicon
- 5xxx – Aluminium–magnesium
- 6xxx – Aluminium–magnesium–silicon
- 7xxx – Aluminium–zinc

In the UNS numbering system, the alloys are denoted by the letter 'A' followed by five numbers. The first digit following the initial letter 'A' defines the type of product. The most commonly used number in this case is 9, which signifies a wrought product. Numbers other than 9 are used to define other products such as castings, ingots, and clad products, among others. The last four numbers are the same as those of the Aluminium Association numbering system listed above.

6.2.2   General properties

For seawater applications, the alloys of choice are the 5xxx-series alloys because of their superior corrosion resistance. Some of the 6xxx alloys are also used, but alloys of the other series, particularly those containing copper, show inferior corrosion performance in seawater.

In addition to the alloy number designation, a further designation denoting the temper condition of the alloy is used. This takes the form of a suffix following the alloy number and gives the details of how the material was strengthened. For non-heat-treatable alloys, this is denoted by Hxxx (e.g. 5083-H321). For heat-treatable alloys, it is designated by Txxx (e.g. 6082-T6). If no hardening is done, or the product is fully annealed, then the temper is given by -O (e.g. 5086-O).

Aluminium alloys can be produced as castings or as wrought (hot- and/or cold-worked) products. Because of the significant differences between these two product forms, their numbering system is separate. In as-cast products, the as-cast micro-structure normally has low strength and ductility, whereas wrought products offer a recrystallised microstructure (although of a directional nature) giving improved mechanical and engineering properties. Hot-working methods usually take the form of rolling, extrusion, or forging. Sheets of thicknesses below ~2.5 mm are produced by cold rolling.

6.3   Aluminium alloys for seawater applications

Aluminium alloys are favoured for uses for which their high strength-to-weight ratio, as well as a high corrosion resistance, are particular advantages [1,2]. Table 6.1 provides examples of where the higher corrosion-resistant aluminium alloys are used in marine environments. Table 6.2 and Table 6.3 give chemical compositions and mechanical properties, respectively, for many of the aluminium alloys used in seawater service.

Table 6.1 Typical aluminium alloy applications in marine environments

Alloy type	UNS no.	Applications
5083/5086	UNS A95083/ UNS A95086	Yachts, fishing boats, pleasure craft, customs vessels, and police boats
5086	UNS A95086	Liquefied natural gas (LNG) storage containers
6005A/6061	UNS A96005/ UNS A96061	Gangways, pontoons, catwalks, ladders, and helidecks
5086/6082	UNS A95086/ UNS A96082	Fin tube heat exchangers

Table 6.2 Chemical composition of the main wrought aluminium alloys used in seawater

UNS no.	Nominal composition, wt%								
	Mg	Zn	Si	Fe	Mn	Cu	Cr	Ti	Zr
A95083	4.5	0.25 max	0.40 max	0.40 max	0.7	0.10 max	0.15	0.15 max	–
A95086	4.0	0.25 max	0.40 max	0.50 max	0.4	0.10 max	0.15	0.15 max	–
A95754	3.0	0.20 max	0.40 max	0.40 max	0.50 max	0.10 max	0.30 max	0.15 max	–
A95456	5.0	0.25 max	0.25 max	0.40 max	0.70	0.10 max	0.1	0.20 max	–
A95059	5.5	0.6	0.45 max	0.50 max	0.9	0.25 max	0.25 max	0.20 max	0.15
A95383	4.5	0.25 max	0.25 max	0.25 max	0.8	0.10 max	0.15	0.15 max	0.2 max
A96005	0.5	0.20 max	0.7	0.35 max	0.50 max	0.30 max	0.30 max	0.10 max	–
A96061	1.0	0.25 max	0.6	0.70 max	0.15 max	0.3	0.2	0.15 max	–
A96082	0.9	0.20 max	1.0	0.50 max	0.7	0.10 max	0.25 max	0.10 max	–

6.4   Main corrosion types

Aluminium alloys in normal use corrode either by a mechanism involving general attack, or by a localised mechanism [3]. In seawater, provided the natural oxide film is maintained, general corrosion is rarely seen, as this type of corrosion mainly occurs when the environment pH is lower than 4 or greater than 8.5. Localised corrosion occurs in seawater because of a breakdown of the protective oxide film in isolated areas, and results in pits developing on the surface. The initiation of such pitting is a chemical attack by chloride ions at defective points in the passive oxide film [4].

Of the aluminium alloys available, the 5xxx alloys possess favourably high corrosion resistance in seawater [3,5]. Generally, the 5xxx alloys are less prone to pitting corrosion than the 6xxx alloys. The presence of copper in aluminium alloys, such as UNS A96061 (6061), produces deleterious cathodic defects in the oxide film. The detrimental effect caused by the presence of copper becomes even greater for 2xxx (Al–Cu) alloys; therefore, these alloys are not suitable for seawater applications. The high-strength 7xxx (Al–Zn) alloys also suffer from excessive corrosion in seawater and are rarely used in marine applications.

6.5   Other types of corrosion

6.5.1   Crevice corrosion

Crevice corrosion is a particular form of localised corrosion and occurs when a crevice in a metallic joint allows a high concentration of chloride ions to develop, resulting in a breakdown of the passive oxide film. Examples of common crevice types are those found underneath a bolt head, gasket, or washer, or between a shaft and bearing. The occluded nature of the crevice allows low-pH environments to develop, which locally speed up the corrosion process. Crevice-type corrosion can also be stimulated by the presence of wet, porous insulating materials when they are in contact with aluminium surfaces.

6.5.2   Waterline corrosion

Waterline corrosion occurs just below the air–seawater interface on aluminium structures that are semi-submerged in seawater [1]. It is more prevalent in stagnant

Table 6.3 Normal form and mechanical properties of the main aluminium alloys used in seawater with various conditions of temper

Alloy	Product form	Alloy temper and condition	0.2% Proof stress, N mm^{-2}	Tensile strength, N mm^{-2}	Elonga-tion on 5.65 S_0, %
A95083	Plate/Sheet	O/H111	125	275	15
		H112	125	275	10
		H116	215	305	10
		H321	215	305	10
		Welded	125	275	N/A
A95086	Plate/Sheet	O/H111	100	240	16
		H112	125	250	9
		H116	195	275	9
		H321	195	275	10
		Welded	100	240	N/A
A95754	Plate/Sheet	O/H111	80	190	17
A95059	Plate/Sheet	O/H111	160	330	24
		H116	270	370	10
		H321	270	370	10
		Welded	160	300	N/A
A95456		O	130	290	16
		H116	230	315	10
		H321	230	315	12
A95383	Plate/Sheet	O/H111	145	290	17
		H116	220	305	10
		H321	220	305	10
		Welded	145	290	N/A
A96061	Plate/Sheet	T5/T6	240	290	10
		Welded	125	160	N/A
A96082	Plate/Sheet	T5/T6	240	280	8
		Welded	125	190	N/A
A96005-A	Extruded	Open profile T5/T6	215	260	6
		Open profile T5/T6 Welded	100	160	N/A
		Closed profile T5/T6	215	250	5
		Welded	100	160	N/A
A96061	Extruded	Open profile T5/T6	240	260	8
		Open profile T5/T6 Welded	125	160	N/A
		Closed profile T5/T6	205	245	4
		Welded	125	160	N/A
A96082	Extruded	Open profile T5/T6	260	310	8
		Open profile T5/T6 Welded	125	190	N/A
		Closed profile T5/T6	240	290	5
		Welded	125	190	N/A

conditions than those where the seawater is in motion. It is caused by the development of differences in chloride ion concentration when chloride ions become concentrated through evaporation at a stationary meniscus. A method of avoidance is to coat the area on either side of the air/seawater boundary.

6.5.3   Microbiologically influenced corrosion (MIC)

Certain microorganisms can become attached to aluminium alloys from seawater or from contaminated fuel and give rise to slimy biofilms on the surface. These films can accelerate the initiation of pitting corrosion. Normal preventative measures are the use of coatings and biocides. Biofilm activity usually ceases at a temperature of 30 °C above the normal ambient temperature.

6.5.4   Galvanic corrosion

Aluminium is a very reactive metal, and when placed in contact with another metal in the presence of moisture, an electrolytic cell is set up [6]. This situation could result in preferential attack of the aluminium, an effect known as galvanic corrosion. The driving force behind galvanic corrosion is the electrochemical activity difference between the two metals coupled together. Aluminium and its alloys are very electro-chemically active, and when exposed to an electrically conducting medium (e.g. seawater), they tend to be the ones that corrode preferentially if any other metal is in contact. Carbon in contact with aluminium, most commonly found when incorpo-rated in gasket materials, readily forms a galvanic cell and causes the aluminium to be preferentially attacked.

Normal methods of galvanic corrosion prevention are used, such as coating the surface of the more noble material (the cathode), electrically isolating the two metals, or applying cathodic protection (CP) (see Chapter 8 for more on galvanic corrosion). The latter must be carefully designed because locally high pH values can be generated on aluminium at cathodic potentials, which can produce a breakdown of the passive film.

6.6   Intergranular corrosion

Intergranular corrosion is a form of galvanic corrosion that takes place on a micro scale [7]. Some aluminium alloys can form almost continuous layers of very active intermetallic phases at grain boundaries if they experience certain thermal exposures (e.g. if non-ideal hot-working methods are used). UNS A95083 (5083) is susceptible to this phenomenon, as precipitates of the $MgAl_3$ phase can form at the grain bound-aries when the material is exposed to a temperature range of 65 to 200 °C.

Immersion of UNS A95083 (5083) with severe grain boundary precipitation in a corrosive medium results in preferential attack of the material at the grain boundar-ies, producing intergranular corrosion. Thus, the maximum service temperature for UNS A95083 (5083) is restricted to 65 °C.

Generally, 5xxx-series alloys with magnesium content lower than 3.5% are con-sidered immune to intergranular corrosion. The 6xxx-series alloys do not normally display a susceptibility to intergranular corrosion.

6.6.1   Exfoliation corrosion

Exfoliation corrosion is a specific form of intergranular corrosion that occurs on products manufactured in such a way as to make the grain boundaries prone to preferential attack [8]. Rolled or extruded materials are sometimes manufactured in such a way that the production process allows electrochemically active precipitates to form along the grain boundaries.

In this type of corrosion, intergranular attack progresses along the grain boundaries and the corrosion product, which has a higher relative volume than the original material, forces the grains apart, causing them to flake off. Because the resulting corrosion gives the separated grains an appearance similar to that of piles of leaves, the corrosion type has become known as exfoliation. This form of corrosion is particularly aggressive and causes a rapid decrease in cross-sectional area. In general, the only repair strategy involves cutting out the affected area.

Exfoliation corrosion is avoided by making sure manufacturing methods are fully controlled and product samples are tested before release. Standard accelerated test methods are available that stimulate exfoliation corrosion if a susceptible microstructure is present. ASTM G66 [9] and ASTM G67 [10] are the two most widely recognised test procedures.

6.6.2   Stress corrosion cracking

Stress corrosion cracking (SCC) can occur in some 5xxx- and 7xxx-series aluminium alloys if the alloy has a susceptible microstructure, particularly in cases in which the manufacturing process or subsequent thermal and/or mechanical exposure has allowed electrochemically active precipitates to form at the grain boundaries. Residual stresses in the material provide the required level of stress for cracking to occur. The 6xxx-series alloys are considered immune to SCC.

6.6.3   Corrosion fatigue

Under cyclical loading conditions in the presence of seawater, the fatigue strength at 10^8 cycles of all aluminium alloys is normally reduced from that found in air by 25% to 35%. For a particular alloy, the corrosion fatigue performance is virtually independent of its metallurgical condition.

6.7   Design considerations

The design of an engineering structure has a significant influence on its corrosion resistance. Therefore, design considerations can be made to reduce the likelihood of aluminium alloy corrosion in seawater. Some of these are as follows:

- Avoid contact between dissimilar metals as galvanic corrosion is a common cause of corrosion in aluminium structures. If contact is unavoidable, the surfaces should be electrically isolated from each other.
- Avoid crevices as these can be present in the design, but also occur as a result of weld defects, for example, where there are undercuts, lack of fusion, or partial penetration features. In areas where crevices cannot be avoided, the ingress of moisture should be prevented by applying coatings or sealants.
- Use continuous welding where possible. A continuous weld acts as a physical barrier to water ingress. Intermittent welding and riveting also result in the formation of crevices.
- Use good drainage systems. Corrosion can occur in stagnant water conditions. This is enhanced when water evaporates from the undrained seawater, increasing the chloride concentration. Pitting readily occurs under such conditions in all aluminium alloys.

- Avoid sharp bends in pipes, as these can result in erosion–corrosion of aluminium. The constant removal of the surface does not allow a protective oxide to form, and corrosion rates are enhanced.
- Avoid excessive stress concentrations, which can cause SCC and an increased rate of corrosion fatigue.
- Avoid sharp edges, which are an example of details that are difficult to coat in an effective manner. Components that are coated should be designed to provide the best coating condition.

6.8   Organic coatings

Although bare aluminium alloy surfaces can develop a good protective patina after exposure to seawater, organic coatings can be applied to produce a physical barrier between the aluminium and its environment. The effectiveness of a coating depends as much on surface preparation as on coating selection. A poorly prepared surface causes the failure of even a good coating. There are a number of pretreatments that can be applied, the choice of which depends on the coating system used. The normal stages of coating application for aluminium are as follows.

6.8.1   Surface cleaning

When preparing an aluminium surface for coating, the first step is to remove all corrosion and corrosive material from the surface. This is best achieved with high-pressure, freshwater cleaning with an applied pressure of at least 17.2 N mm^{-1} [2]. The run-off water should flow free from the aluminium.

Next, a method of abrasion is applied that is designed to produce a surface with an oxide film that is as thin and uniform in thickness as possible. The normal method of achieving this is through the use of grit blasting with non-metal abrasives, or abrasion with 60- to 120-grade abrasive paper.

6.8.2   Pretreatment

Bare metal surfaces must be primed immediately after surface preparation to avoid too much oxide thickening, contamination, or general surface deterioration, and to allow maximum adhesion to the surface. A pretreatment conversion coating is applied, which converts the aluminium oxide at the surface into a layer that essentially inhibits the corrosion process. Historically, chromium compounds were used for this, but environmental considerations have led to the development of non-chromium primers that are now widely available.

6.8.3   Primers

Primers are usually epoxy-based and they are conversion coatings applied in multiple coats following pretreatment. The time between coats and the recommended coating thicknesses for different applications are provided in the information documents for the individual primer products.

6.8.4   Top coats

The technology of organic coating is complex and in a state of constant development. The success of any coating system depends on the pretreatment. The most common topcoats are polyester-based, and are generally chosen for corrosion protection in marine environments because of the flexibility of the coating and the relatively low cost.

6.8.5   Corrosion inhibitors

Typical corrosion inhibitors for aluminium are chromates, phosphates, silicates, nitrates, and benzoates. However, the application of inhibitors in marine applications is most successful when the inhibitors are used in enclosed seawater systems or they are bound to coatings. When inhibitors are added to seawater, reliable systems need to be in place to monitor their concentration to ensure they remain effective.

6.8.6   Cathodic protection

The corrosion of aluminium may be prevented by reducing the corrosion potential to a level at which the metal is immune to corrosion [1]. The current methods of achieving such immunity may come through the application of zinc or aluminium indium sacrificial anodes, or by an impressed current system. The design of the CP system must be such that the potentials applied do not allow high local pH values that could permit passive film breakdown to occur.

6.9   Joining methods

6.9.1   Welding

In general, the wrought alloys comprising the 1xxx-, 3xxx-, 4xxx-, 5xxx-, and 6xxx-series can be welded by the fusion-welding process. However, the strength properties of the heat-affected zone (HAZ) produced are normally lower than that of the parent material. For work-hardening alloys, the strength of the HAZ is equivalent to the annealed condition (-O).

The main significance of welding with respect to the corrosion performance of aluminium alloys is geometrical, as described earlier in this chapter. Thus, the introduction of crevices, for example, is probably more significant than the metallurgy of the welds. Consequently, the interesting galvanic effects introduced by welding heat-treatable alloys, either because of the thermal cycle in the HAZ or as a result of the use of dissimilar composition fillers in the weld metal, are of minor importance to the alloys that are typically used.

Non-matching filler metals are often used for welding aluminium, principally to avoid hot-cracking problems. For example, 6xxx alloys can be welded with either 4xxx or 5xxx alloys to prevent solidification cracking. Al–Si weld metals are cathodic to the parent metal, whereas weld metals of alloy type Al–Mg are anodic. Thus, for seawater service, it is preferable to use Al–Si consumables for joining 6xxx alloys.

The 5xxx-series alloys with sufficient magnesium can be sensitised to intergranular corrosion or SCC by specific thermal exposures; however, welding thermal cycles do not cause sensitisation.

The introduction of friction-stir welding has led to the application of welding to a wider range of aluminium alloys but not to their service in seawater.

6.9.2   Joining aluminium to steel structures

Welding of aluminium directly to steel by fusion welding is not possible as brittle iron/aluminium intermetallics are formed that give the joint a high susceptibility for cracking. However, aluminium and steel can be closely joined by roll bonding, explosion bonding, and resistance or friction welding. Therefore, it is possible to produce aluminium/steel transition joints. These are used to join aluminium and steel structures using a process that involves welding the aluminium part of the structure to the aluminium side of the transition joint, and the steel structure to the steel side of the joint.

However, with this method, it is difficult to achieve the full material strength because of the reduced bond strength between the clad steel and the aluminium alloy. Typical values for the bond strength are 70 N mm^2 in tension and 55 N mm^2 in shear. Care is required when making the fillet welds – the maximum temperature in the bond area must be kept to a minimum to prevent loss of cohesion. It is usual to limit this temperature in accordance with the clad component manufacturer's recommendations, and the maximum temperature is usually listed in the range of 250 to 350 °C.

A disadvantage of transition joints is an enhanced susceptibility to corrosion by galvanic attack. Therefore, the assembly should be coated to provide adequate protection.

6.10   Other aluminium joining techniques

Joints between aluminium alloys and between aluminium and other materials can be made using rivets, screws, bolts, and adhesive bonding. For rivets, it is advisable to use Al–Mg (5xxx-series) alloys, particularly UNS A95754 (5754), as these materials are less susceptible to SCC than other aluminium alloys. The choice of materials for screws and bolts depends on whether the joint is submerged in seawater or not. If the joint is above water, then stainless steel, galvanised steel, or cadmium-plated steel can be used. If submerged, mild steel or stainless steel cannot be used without some form of protection. This could be CP or electrical isolation placed between the aluminium and the fastener. Aluminium fasteners should not be used for joining metals other than aluminium. The large area ratio that would result in such cases would be likely to cause rapid corrosion of the fasteners.

If adhesive bonding is used, then good surface preparation is vital. This should consist of a thorough degreasing operation followed by the application of a conversion coating or wash primer, both of which chemically condition the surface so its bonding performance is optimised.

Acknowledgements

The author would like to acknowledge the valuable discussion with colleagues in the Materials Department of Lloyd's Register on a number of subjects included in the chapter. Thanks should also be expressed to Stuart Bond of TWI for contributing the section on welding and to Tom Siddle and Will Savage of the Aluminium Federation for their helpful suggestions during the preparation of the final text.

References

1. C. Vargel: 'Corrosion of aluminium'; 2004, Amsterdam, Netherlands, Elsevier BV, 2004.
2. J. R. Davis, ed.: 'ASM specialty handbook: Aluminum and aluminum alloys', 579–622; 1993, Materials Park, OH, ASM International.

3. C. Vargel: 'Aluminium and the sea'; 1993, Montreal, Quebec, Canada, Rio Tinto Alcan [formerly ALCAN].
4. M. C. Reboul *et al.*: *Mater. Sci. Forum*, 1996, **217/222**, 1553–1558.
5. W. H. Ailor, Jr. *et al.*, ed.: in 'Corrosion in natural environments', STP 558, West Conshohocken, PA, ASTM, 1974, 117.
6. M. C. Reboul: *Corrosion*, 1979, **35**, (9), 423; and *Rev. Aluminium*, 1977, **465**, 404–419.
7. H. Kaesche: in 'Localised corrosion', (ed. B. F. Brown, J. Kruger and R. W. Staehle), Vol. 3, 516, International Corrosion Conference, Williamsburg, VA, 1974, NACE International, Houston, TX.
8. J. Zahavi and J. Yaholom: *J. Electrochem. Soc.*, 1982, **129**, (6), 1181–1189.
9. ASTM: 'Standard test method for visual assessment of exfoliation corrosion susceptibility of 5XXX series aluminum alloys (ASSET test)', ASTM G66, ASTM, West Conshohocken, PA.
10. ASTM: 'Standard test method for determining the susceptibility to intergranular corrosion of 5XXX series aluminum alloys by mass loss after exposure to nitric acid (NAMLT test)', ASTM G67 (latest revision), ASTM, West Conshohocken, PA.

7
Titanium alloys

Ronald W. Schutz

RTI International Metals, Inc., 1000 Warren Avenue, Niles, Ohio 44446, USA

rschutz@rtiintl.com

7.1 Introduction

Titanium and its alloys are traditional materials used in marine service, such as seawater cooling systems [1–7], desalination systems [1,5], naval ship components [1–3,6,8–11], and offshore hydrocarbon production systems [1,2,4,6,7,12–15] based on their virtual immunity to seawater corrosion and/or their high strength-to-weight ratio. While they are more costly materials compared to most stainless steels, their use becomes justifiable for more severe, critical service applications in which maintenance and life-cycle costs are substantial, downtime is costly or a non-option, and/or enhanced structural efficiency (i.e. weight reduction) or specific performance features are critical.

Successful, cost-effective use of titanium and its alloys is predicated on understanding its unique combination and range of mechanical, physical, and corrosion properties compared to classic marine engineering alloys to take full performance advantage in the design of components. To address these materials, this chapter presents an overview on relevant titanium alloys, their traditional marine applications, and seawater corrosion performance, while offering guidance on avoiding cathodic protection (CP) system hydrogen damage and achieving galvanic compatibility in seawater components and systems.

7.2 Characteristics of titanium and its alloys

Titanium is represented by a family of commercial alloys with a very wide range of strength (Table 7.1), unique physical properties compared to conventional metals, and exceptional resistance to corrosion and erosion in seawater and other aqueous chloride media. Those alloys designated with the superscript '(A)' represent those more commonly and traditionally used in seawater and marine applications. Selection of titanium alloys for seawater applications is normally predicated on the expected practical immunity to general and/or localised corrosion in seawater (i.e. a zero corrosion allowance), but may also be driven by a combination of often synergistic design features depending on the component application. An overview of the diverse characteristics of titanium alloys for application and design consideration is provided in Table 7.2.

The titanium alloys traditionally used in seawater/marine service, listed in Table 7.1, are readily welded, machined, and fabricated, with some adjustments, using similar methods and equipment used for stainless steels. Standard mill product forms for these common marine titanium grades are commercially available, and typically

Table 7.1 Common commercial titanium alloys considered for seawater/marine service

Common name	ASTM grade	UNS no.	Nominal composition, wt%	Alloy type	Min. 0.2% YS, MPa
CP-Grade 1[A]	1	R50250	Unalloyed titanium	Alpha	172
CP-Grade 2[A]	2	R50400	Unalloyed titanium	Alpha	275
CP-Grade 3[A]	3	R50550	Unalloyed titanium	Alpha	380
Ti-Pd[A]	7	R52400	Ti–0.15Pd	Alpha	275
'Soft' Ti-Pd[A]	11	R52250	Ti–0.15Pd	Alpha	172
Ti-lean Pd	16	R52402	Ti–0.05Pd	Alpha	275
Ti-Ru	26	R52404	Ti–0.1Ru	Alpha	275
Grade 12[A]	12	R53400	Ti–0.3Mo–0.8Ni	Alpha	345
Ti-3-2.5[A]	9	R56320	Ti–3Al–2.5V	Near-alpha	483
Ti-3-2.5-Pd	18	R56322	Ti–3Al–2.5V–0.05Pd	Near-alpha	483
Ti-3-2.5-Ru	28	R56323	Ti–3Al–2.5V–0.1Ru	Near-alpha	483
Ti-5111[A]	32	R55111	Ti–5Al–1Sn–1Zr–1V–0.8Mo	Near-alpha	586
Ti-6-4[A]	5	R56400	Ti–6Al–4V	Alpha beta	827
Ti-6-4-ELI[A]	23	R56407	Ti–6Al–4V (0.13% O max)	Alpha beta	759
Ti-6-4-Ru[A]	29	R56404	Ti–6Al–4V–0.1Ru (0.13% O max)	Alpha beta	759
Ti-38-6-44	19	R58640	Ti–3Al–8V–6Cr–4Zr–4Mo	Beta	1103

[A]Alloys more commonly and traditionally used in seawater and marine applications.

specified via relevant ASTM product specifications that provide composition and tensile property requirements.

7.3   Titanium alloys in marine use

The wide range of titanium alloys that are used traditionally are grouped into five categories as alpha, near-alpha, alpha beta, near-beta and beta titanium [16,17]. Those that are candidates for specialised applications are included in Table 7.1. The alpha alloys represent the lower-strength, unalloyed (or modified unalloyed) titanium grades, from which the commercially pure (unalloyed) UNS R50250 (Grade 1) and, especially, UNS R50400 (Grade 2) are used extensively in seawater cooling systems (heat exchangers and piping), primarily for corrosion resistance [1–7,11]. The near-alpha alloys offer medium strength, high fracture toughness, and ease of fabrication for weldable structural and/or pressurised components.

Increasing the beta alloying content produces the alpha beta titanium alloys that offer high strength (≥759 MPa YS), from which the versatile, weldable Ti–6Al–4V-based alloys (UNS R56400 [Grade 5], UNS R56400 [Grade 23], and UNS R56401 [Grade 29]) have found substantial use in marine service. These two-phase alloys offer the ability to achieve a more favourable balance in strength, ductility, and fracture resistance through microstructural variations and tailored wrought process-ing. The Ti–6Al–4V-based alloys also are available in extra-low interstitial (ELI) 0.13 wt% maximum oxygen form that is preferred when seawater fracture toughness or fatigue crack growth resistance is critical [18]. The most highly alloyed titanium near-beta and beta alloys offer very attractive elevated strength (≥1103 MPa YS) when aged, but have not enjoyed substantial marine service use to date because of their higher cost and diminished machinability and weldability.

Table 7.2 Diverse/synergistic characteristics of titanium alloys for component/
equipment design in seawater

Characteristic	Use/Design implication
Essentially immune to seawater corrosion (wrought, weld, or cast forms) Low density (half of Ni and Cu alloys) Elevated strength-to-density ratio	Zero corrosion allowance and thinner wall sections Lightweight, weight reduction Lightweight, high structural efficiency, and reduced centrifugal forces
Elevated erosion/erosion–corrosion/ cavitation resistance	Thinner wall sections, lighter weight, much higher flow rates, and turbulent flow design possible. Good for pumps, valves, propellers, heat exchangers, and piping
Resistant to corrosion fatigue	No reduction in S-N fatigue life in seawater. Good for dynamic offshore risers and rotating components (pump impellers, shafts, and propellers)
Roughly half the elastic modulus of steel	Designed with lower stiffness in mind – good for deflection-controlled components (risers and springs), but may need to reinforce other structures and modify heat exchanger tube support spans
Unalloyed Ti thermal conductivity ~30% higher than austenitic SS	Good heat transfer and size reduction in heat exchangers
Not biotoxic, environmentally friendly, and immune to MIC	Surfaces can macro- or micro-biofoul, but biocorrosion is not possible. Can use chlorination for antifouling treatment
Elevated shock and ballistic resistance	Military/naval vessel components for survivability (e.g. lightweight armour, hulls, hatches, and doors)
Essentially nonmagnetic	Navy mine hunter vessel components
Galvanically compatible with carbon/ carbon and graphite reinforced composites	Design structurally efficient Ti/composite hybrid structures
Susceptible to galling and sliding metal wear	Use appropriate antigalling/lubricant surface treatments for fasteners, threaded connections, and bearing surfaces
Most Ti grades are readily weldable to themselves, but not to dissimilar metals	Use mechanical connection methods when joining Ti to other metals. Design to preclude galvanic corrosion, if required

MIC, microbiologically influenced corrosion.

Table 7.3 provides an overview of documented applications and specific components for titanium alloys used in seawater environments. The primary use for titanium in these arenas involves UNS R50250 (Grade 1) and UNS R50400 (Grade 2) used in plate/frame and shell/tube seawater coolers/condensers to fully resist the seawater coolant [5–11]. A relevant example is the more than 198 million metres of seam-welded UNS R50400 (Grade 2) titanium tubing used in seawater-cooled steam surface condensers of coastal power plants worldwide that have never experienced a seawater corrosion-related failure during 40 years of application. For structural,

Table 7.3 Titanium alloy applications in seawater service

Arena	Ti alloy UNS number (ASTM Ti Grade)	Specific seawater applications
Coastal processing operations (oil refineries, chemical and petrochemical plants)	UNS R50400 (Grade 2)/ UNS R52400 (Grade 7)/ UNS R53400 (Grade 12)/ UNS R52402 (Grade 16)	Shell/tube coolers and condensers
	UNS R50250 (Grade 1)/ UNS R52250 (Grade 11)	Plate/frame coolers
	UNS R50400 (Grade 2)	Seawater piping
Power plants	UNS R50400 (Grade 2)/ UNS R50550 (Grade 3)	Steam surface condensers (shell/tube)
	UNS R50250 (Grade 1)/ UNS R50400 (Grade 2)	Auxiliary coolers (shell/tube and plate/ frame)
	UNS R50400 (Grade 2)	Cooling water piping
	UNS R50250 (Grade 1)/ UNS R50400 (Grade 2)	Electrolytic chlorinators Offshore thermal energy conversion
	UNS R50400 (Grade 2)	Ocean thermal-energy conversion (OTEC) shell/tube boilers and condensers
Desalination plants	UNS R50250 (Grade 1)/ UNS R50400 (Grade 2)/ UNS R53400 (Grade 12)	Shell/tube and plate/frame brine heaters/ preheaters, heat rejection exchangers, evaporator heaters, and final condenser
Offshore hydrocarbon production	UNS R56400 (Grade 23)/ UNS R56401 (Grade 29)/ UNS R56320 (Grade 9)	Offshore production riser taper stress joints, drilling risers, and drill mud booster lines
	UNS R50250 (Grade 1)/ UNS R50400 (Grade 2)/ UNS R52400 (Grade 7)/ UNS R53400 (Grade 12)	Gas/oil product coolers and condensers (shell/tube and plate/frame)
	UNS R50250 (Grade 1)/ UNS R50400 (Grade 2)	Various topside coolers (lube-oil, engine, compressor, glycol, and others)
	UNS R50400 (Grade 2)/ UNS R56400 (Grade 5)	Topside seawater cooling and fire main piping, and pumps
	UNS R50400 (Grade 2)/ UNS R56400 (Grade 5)	Ballast water system piping and pumps
	UNS R50400 (Grade 2)	Seawater lift pipes
	UNS R50400 (Grade 2)	Electrolytic chlorinator vessels, valves, and piping
	UNS R50400 (Grade 2)	Cable support fittings and fixtures for mooring/station-keeping systems
	UNS R56400 (Grade 23)	Hulls/framework/pressure vessels for manned submersibles
Naval surface ships/submarines	UNS R50250 (Grade 1)/ UNS R50400 (Grade 2)	Shell/tube and plate/frame electronic coolers
	UNS R50250 (Grade 1)/ UNS R50400 (Grade 2)	Various shell/tube and plate/frame coolers (lube-oil, engine, weapons, low- and high-pressure air conditioning [AC] coolers), and condensers (refrigeration/AC)

Table 7.3 Continued

Arena	Ti alloy UNS number (ASTM Ti Grade)	Specific seawater applications
Naval Surface ships/submarines continued	UNS R50400 (Grade 2)/ UNS R50550 (Grade 3)	Submarine main steam condensers (shell/tube)
	UNS R50400 (Grade 2)	Seawater service and fire main piping system
	UNS R50400 (Grade 2)	Sewage/collection, holding, and transfer (CHT) piping
	UNS R50400 (Grade 2)/ UNS R56400 (Grade 5)	Valves, pumps, fire pumps
	UNS R50250 (Grade 1)/ UNS R50400 (Grade 2)/ UNS R50550 (Grade 3)	Water purification systems (distillation and reverse osmosis)
	UNS R55111 (Grade 32)	Surface ship masts
	UNS R56400 (Grade 23)/ UNS R55111 (Grade 32)	Fasteners
	UNS R56320 (Grade 9)	Exhaust uptake liners
	UNS R50250 (Grade 1)/ UNS R50400 (Grade 2)	Recessed electrical boxes and light fixtures
	UNS R50400 (Grade 2)/ UNS R56400 (Grade 5)	Hydrofoil ship components (propulsion impellers, fasteners, and struts)
	UNS R50550 (Grade 3)/ UNS R56320 (Grade 9)/ UNS R56407 (Grade 23)	Deep diving submersibles and research vehicles (buoyancy spheres, pressure vessels, framework, and piping)
	UNS R56320 (Grade 9)	Deep-diving submarine pressure hulls
	UNS R56407 (Grade 23)	Submarine pressure hull doors/torpedo bay doors
	UNS R58640 (Grade 19)	Aircraft carrier water brake cylinders
	UNS R56400 (Grade 5)	Marine gas turbines for ships (compressor blades and discs)
	UNS R50400 (Grade 2)/ UNS R56400 (Grade 5)	Submarine sonar masses/plates
Coastal infrastructure	UNS R50250 (Grade 1)/ UNS R50400 (Grade 2)	Coastal building roofing and siding/ cladding. Bridge and pier piling sheathing

highly loaded, and/or pressurised static or dynamic marine components, the use of UNS R56400 (Grade 23) and UNS R56401 (Grade 29) for various risers in offshore hydrocarbon production [4,7,12–14], and long-time use of UNS R56400 (Grade 5) and UNS R56400 (Grade 23) in various naval ships and submersibles are noteworthy (Table 7.3) [1,6,8,9].

7.4   Corrosion resistance

The exceptional corrosion resistance of titanium alloys in marine/seawater environments originates from the formation of a thin (typically 5 to 30 nm thick) highly

chemically stable, adherent, ceramic-like protective surface oxide film. This tenacious oxide film is primarily TiO_2, and forms spontaneously and instantaneously when fresh metal surfaces are exposed to mere traces of oxygen (air) or water (moisture) alone. As such, a mechanically damaged (e.g. scratched) oxide film can instantly repair/heal itself in air or submerged in aerated or fully deaerated seawater. The stability/formation of this TiO_2 film is not inhibited/affected by the presence of chlorides, and can occur over a wide range of pH values depending on temperature (i.e. pH 2 to 12 at room temperature) [19]. For natural seawater exposures, the TiO_2 passive film is stable at any temperature, and is not affected by conditions that may alter the corrosion resistance of other alloy systems, such as adjoining component CP systems, biofilm formation, or seawater chlorination.

As such, titanium alloys are fully resistant to ambient or near-ambient temperature seawater in aerated, deaerated, fully chlorinated, or even contaminated conditions. The presence of metal ions, sulphides, organics, decaying organics, CO_2, and/or ammonia in contaminated seawater has no effect on corrosion resistance. Essentially nil general corrosion rates (<0.002 mm year^{-1}) can be expected for all titanium alloys indefinitely exposed in submerged seawater, splash zones, or marine atmospheres, permitting component design to zero corrosion allowance [7,19–21]. This exceptional seawater resistance can be expected in the appropriate wrought, cast, and welded titanium product forms, regardless of metallurgical or heat-treated condition.

7.4.1   Resistance to erosion–corrosion

Titanium's thin, but hard and rehealable surface TiO_2 film also explains why titanium and its alloys exhibit exceptional erosion–corrosion and cavitation resistance in seawater. As such, titanium can resist turbulent seawater flow to velocities in excess of 36 m s^{-1} without significant metal loss, or withstand seawater laden with sand/silt to velocities as high as 4.6 m s^{-1} [19,22]. This permits design and operation of seawater piping and heat exchanger systems to significantly higher flow velocities than traditional copper alloy components, offering a means for mitigating surface biofouling effects and chlorination needs, and enhanced heat transfer for reducing heat exchanger size. With their harder substrates, the higher-strength alpha beta and beta titanium alloys further extend resistance to erosive wear and cavitation [3,19–21] compared to unalloyed titanium and most other marine metals. In this regard, titanium alloys can be optimal materials for pumps, valves, impellers, and propellers used under highly turbulent seawater flow conditions.

7.4.2   Resistance to localised corrosion

Titanium alloys exhibit exceptionally high resistance to pitting in aqueous chloride media including seawater, compared to other common engineering metals. As such, they are not susceptible to spontaneous chloride-induced pitting in seawater to temperatures in excess of 200 °C [7,12,13,19,20,22]. This stems from titanium's stable oxide film and elevated anodic pitting potentials that exceed several volts, even in hot seawater [19,20]. Total pitting resistance can be expected under any chlorinated and/or aerated seawater conditions, or when ennobling, adherent biofilms form on titanium surfaces.

Although titanium alloys resist all forms of localised attack in natural seawater below approximately 75 to 90 °C, proper alloy selection/use in higher-temperature seawater should be predicated on possible crevice corrosion susceptibility [13,18,19].

Table 7.4 Approximate service temperature limits for titanium alloys exposed to naturally aerated seawater

UNS No. (Titanium Grade)	Temperature limit (based on crevice corrosion threshold)
UNS R50250 (Grade 1)	76 to 80 °C
UNS R50400 (Grade 2)	
UNS R50550 (Grade 3)	
UNS R56400 (Grade 5)	
UNS R56320 (Grade 9)	
UNS R56407 (Grade 23)	
UNS R55111 (Grade 32)	
UNS R52400 (Grade 7)	~330 °C
UNS R52250 (Grade 11)	
UNS R52402 (Grade 16)	
UNS R52404 (Grade 26)	
UNS R53400 (Grade 12)	~270 °C
UNS R56322 (Grade 18)	~300 °C
UNS R56323 (Grade 28)	
UNS R56404 (Grade 29)	
UNS R 58640 (Grade 19)	~200 °C

Crevice corrosion in certain titanium alloys in hot seawater should be considered when severe crevices exist in components (i.e. gasketed flange or threaded connections, O-ring seals, and tube-to-tube-sheet mechanically expanded joints), or form under surface salt deposits. Approximate crevice corrosion threshold temperatures for various titanium alloys are included in Table 7.4 [18,19]. Therefore, when seawater-exposed component surface temperatures exceed 75 to 80 °C, the Pd- or Ru-enhanced titanium alloys, UNS R53400 (Ti Grade 12), or Mo-containing alloys listed become preferred selections.

Titanium is generally compatible with the vast majority of gasket materials, including fluoropolymers and graphite-filled polymers, used in seawater connections and seals. When fluoropolymer thermoplastics or elastomers are used, however, it is important to specify and use the virgin (factory-direct) product forms only. The reprocessed/recycled/remelted fluoropolymer materials should be avoided because localised attack of gasketed titanium surfaces from leachable fluorides may occur in seawater service.

Smeared-in (embedded) carbon steel on titanium equipment surfaces is only a concern if sustained seawater exposure is expected and temperatures exceed ~80 °C [19]. Localised smeared-in iron pitting is possible in hot seawater in unalloyed titanium and Ti–6Al–4V alloys, but is not of concern in the more crevice corrosion-resistant Pd- or Ru-enhanced or Mo-containing titanium alloys (see Table 7.4). Surface smeared-in metal of most other metals/alloys (e.g. Cu, stainless steels, Ni alloys) is not considered a concern for titanium alloys in seawater service.

7.4.3 Biocorrosion resistance

Unlike many common seawater alloys, titanium alloys have demonstrated immunity to all forms of microbiologically influenced corrosion (MIC) in all laboratory and natural service exposures [19,22,23]. This resistance extends to all aerobic and anaerobic bacteria species, biofilms, fungi, algae, and even hard/macrofoulers contacting or adhering to titanium surfaces. This unique ability to fully resist biological activity in seawater results from titanium's high crevice corrosion threshold temperature, and elevated anodic pitting resistance over a wide pH range [19]. Because titanium component surfaces are nontoxic and relatively inert, they can experience macro- and micro-fouling in much the same way as stainless steels and nickel alloys.

7.4.4   Seawater fracture toughness

The unalloyed alpha and near-alpha titanium alloys listed in Table 7.1 are considered resistant to chloride SCC in seawater environments. As such, no significant reduction in fracture toughness (i.e. stress-intensity factor in air $[K_{air}]$ ≈ stress-intensity factor in seawater $[K_{seawater}]$) occurs in seawater, regardless of aeration, impressed cathodic potential, or temperature variations [7,18,19].

Although higher-strength alpha beta and beta alloys are also fully resistant to chloride SCC in loaded smooth or notched component configurations [19], some high-strength alloys may exhibit finite reductions in seawater 'K' value (i.e. in a highly loaded precracked situation) depending on composition, microstructure, and processing condition. Seawater fracture toughness in Ti–6Al–4V and other alpha beta alloys is optimised by selecting an ELI grade, and/or by processing to a fully transformed beta (i.e. acicular alpha) microstructure. This permits successful use of UNS R56400 (Grade 23) and UNS R56401 (Grade 29) in fracture-critical seawater applications such as submarine/submersible pressure hulls/bottles and fasteners, and dynamic offshore risers [1,13,18]. The Ru- and Pd-enhanced near-alpha and alpha beta alloys can be expected to provide SCC resistance under higher-temperature seawater service conditions.

7.4.5   Corrosion fatigue resistance

Overall, titanium alloys are highly resistant to corrosion fatigue, such that S-N lives and endurance limits in air and seawater are practically the same [7,18–20]. This resistance applies to smooth or notched configurations in wrought and weld metal, to temperatures exceeding 100 °C [12,13]. In regard to fatigue crack growth (FCG) in seawater, the same titanium alloys that resist seawater SCC also do not tend to exhibit environmentally enhanced FCG rates in seawater. Thus, the alpha and near-alpha titanium alloys listed in Table 7.1, and the ELI titanium UNS R56400 (Grade 23) and UNS R56401 (Grade 29), all offer superior FCG resistance, similar to their behaviour in air.

7.4.6   Galvanic compatibility

As a result of its protective, passive oxide film, titanium and its alloys fall toward the noble (positive) end of the galvanic series when in seawater. Specifically, they exhibit corrosion potentials in seawater similar to those for passive stainless steels and Ni–Cr–Mo alloys [19–21,24]. As such, titanium can be considered galvanically compatible with and acceptable for direct contact with the more seawater-resistant superaustenitic/superduplex stainless steels, and Ni–Cr–Mo alloys such as UNS N10276 (alloy C-276) and UNS N00625 (alloy 625). Although titanium can experience some potential ennoblement from surface biofilm growth in natural seawater, the extent of ennoblement (positive shift) is similar to that experienced by stainless steels and other corrosion resistant alloys (CRAs) [25].

In contrast to almost all other marine engineering metals, titanium and its alloys can be directly mated to highly noble materials including graphite or platinum group metals, in chlorinated or natural seawater. This is a result of titanium's expanded anodic pitting/repassivation potential regime in aqueous chloride media, whereby additional ennoblement further serves to stabilise/maintain its passive oxide film. As such, these alloys are successfully designed and incorporated into graphite-reinforced polymer or carbon/carbon composite structures or components for marine use.

The primary galvanic concern involves direct coupling of titanium to more active, less-resistant metals (e.g. steel, copper, aluminium, or zinc alloys), or to less-resistant stainless steels in the active condition in seawater. In these situations, coupling to titanium (as a cathode) can stimulate accelerated general and/or localised corrosion of the active metal (anode), depending on seawater temperature, oxygen content and

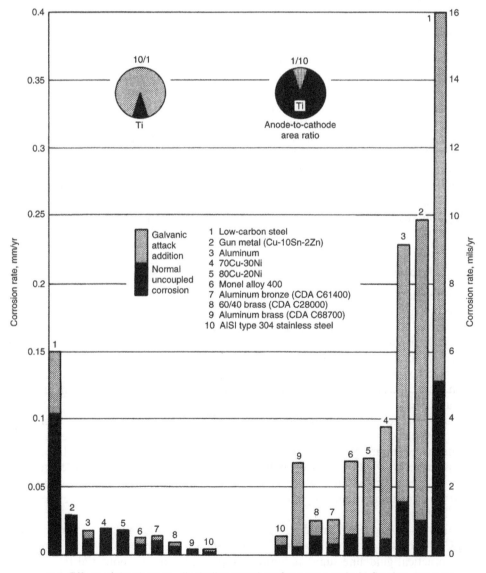

7.1 Effect of titanium-to-dissimilar metal surface area ratio in flowing seawater

flow rate, surface biofilm formation, impressed cathodic potential, and titanium-to-active-metal surface area ratio. This cathode-to-anode area ratio is a particularly significant factor, as indicated in Fig. 7.1 [19,21]. Minimal galvanic corrosion effects involving titanium galvanic couples can be expected when the titanium-to-active-metal surface area ratio is low (i.e. below 0.2) [19–21]. Also, avoiding hydrogen damage in titanium when it is galvanically coupled to highly active metals (i.e. Zn or Al) in seawater should be considered as discussed below in section 7.5 on *Fabrication considerations.*

It should be noted that although titanium has a noble potential in seawater similar to passive stainless steels and high-nickel alloys, it represents a poorer cathode because of its oxide film and associated high cathodic polarisation resistance for oxygen reduction [24–26]. As a result, titanium can be expected to produce somewhat lower galvanic corrosion/currents compared to Ni–Cr–Mo and Ni–Cu alloys when similarly coupled to active metals in sterile seawater. In natural seawater, however, surface biofilms can form over time such that titanium behaves similarly to other CRAs as a cathode in these couples.

Classic strategies for mitigating galvanic corrosion in titanium-to-active-metal couples (and other CRAs) are reviewed in detail in Chapter 8, and include minimising the cathode (Ti)-to-anode surface area ratio (by design or coating the Ti surface), electrically isolating the dissimilar metal components, cathodically protecting the active metal at cathodic potential ranges that avoid excessive Ti hydrogen absorption, or using nonconductive or replaceable piping sections contacting titanium [6,8,13,18,22,26].

Additional considerations may include incorporation of explosively bonded titanium-to-dissimilar-metal transition joints to avoid crevices and simplify galvanic couple interfaces, or applying weld overlays of seawater-resistant Ni–Cr–Mo alloys on steel component faces mated to titanium components (e.g. pipe flange connections).

7.4.7   Hydrogen resistance

Because titanium's surface oxide film serves as a highly effective barrier to both molecular and atomic hydrogen, no significant hydrogen absorption normally occurs in titanium alloys exposed to seawater at any temperature. However, in applications in which titanium is incorporated into seawater components/systems containing other, dissimilar metals requiring CP, consideration should be given to possible cathodic potentials/currents inadvertently impressed on titanium surfaces.

In these cases, impressed-current or sacrificial-anode CP systems typically impress sustained cathodic potentials in the –0.8 to –1.1 V (vs. Ag/Ag-Cl) range that cathodically charges atomic hydrogen onto metal surfaces. Over time, a portion of this atomic hydrogen is absorbed by titanium surfaces, and slowly diffuses into the titanium depending on temperature, alloy type and microstructural condition, and stress state [18,19,22]. Excessive hydrogen absorption can produce brittle titanium hydride precipitate within alpha phase, and induce embrittlement and/or sustained-load cracking in titanium alloy components loaded in tension or under high residual tensile stress.

Effective strategies for avoiding hydrogen damage in titanium alloys from adjoining CP systems depend on the alloy type and functional service involved. A conservative, practical guideline suggests limiting impressed cathodic potential to no more negative than –0.80 V (vs. Ag/Ag-Cl) to preclude hydrogen charging/absorption

[12,18,27]. This should definitely be invoked for long-term exposures involving either higher metal temperatures ($\geq$70 °C), higher-strength near-alpha and alpha beta alloys (including UNS R53400 [Grade 12] titanium), or highly loaded structural components or pressure vessels.

On the other hand, low-stress, lower-temperature service applications such as seawater piping and heat exchanger cooling systems using unalloyed titanium grades may tolerate more cathodic potential limits as negative as −1.05 V (vs. Ag/Ag-Cl). Although a superficial hydride surface film may form, hydrogen penetration is minimal and generally has little practical effect on useful component life. More negative cathodic potentials than this should be avoided as they can cause enhanced hydride penetration even in unalloyed titanium at ambient temperatures.

In certain applications in which these impressed cathodic potential guidelines cannot be practically met, additional design strategies may be invoked to preclude long-term hydrogen damage. These include (1) applying polymeric barrier coatings or sheathing to seawater-exposed titanium surfaces; or (2) electrically isolating (floating) the titanium component from adjoining cathodically protected, dissimilar metal components via isolating flange connections or couplings.

7.5   Fabrication considerations

Unlike stainless steels, titanium alloy components are generally highly insensitive to variations in surface condition in typical marine/seawater exposures. As such, no corrosion concerns exist for titanium components in the pickled, ground, machined, peened, grit-blasted, or cold- or plastically worked surface conditions. Although typically not a problem encountered with qualified fabrication practice, smeared-in (embedded) surface iron or carbon steel is only a potential localised corrosion concern in hot (>80 °C) seawater service with unalloyed titanium and Ti–6Al–4V. The more crevice-resistant grades such as the Pd- or Ru-enhanced and the Mo-containing alloys are generally resistant to this rare higher-temperature seawater phenomenon. Other surface films, coatings, and/or residues including heat-tint (thermal oxide films), paint coatings, organic residues, and biofilms also have no detrimental influence on titanium's seawater resistance.

Although titanium mill products are normally surface-conditioned and are free of any surface alpha case (i.e. a brittle, diffused-in oxygen layer), any subsequent fabrication steps in air involving temperatures exceeding approximately 540 °C, for example, hot forming in air or post-weld stress-relief annealing, should consider whether oxide scale and alpha case removal is necessary. Generally, the depth of alpha case formed is a function of time at temperature, and is generally not a concern for most applications when it is less than 3 to 5 μm deep. Total surface alpha case removal should be done in dynamic, cyclically loaded components in which S-N fatigue life is critical, or with high-strength alloy structural components.

The titanium alloys commonly used in marine service are routinely welded using conventional fusion methods, such as gas tungsten arc (GTAW), plasma, laser, and electron beam (EB) welding. The two key aspects for successful titanium welding are avoiding contamination by ensuring (1) adequate inert gas shielding of face and root sides for molten and heat-affected metal surfaces above approximately 260 °C, and (2) good joint surface cleanliness before welding. Guidelines for welding titanium are well established and can be readily accomplished in most stainless steel welding shops with minor adjustments and training [16,17,28].

Other than achieving contamination-free, ductile welds, titanium welding is a fairly robust process and tolerant to minor variations in welding parameters to achieve acceptable properties. Although postweld stress-relief annealing is normally unnecessary for lower-strength titanium grades with yield strengths below approximately 585 MPa, it is recommended for higher-strength alloys typically used in structural or dynamically loaded applications. Titanium alloys are also very suitable to most solid-state welding methods (e.g. diffusion bonding, friction techniques, and resistance-upset).

Titanium can be used in vessels and heat exchanger tube sheets as solid-wall, loose-liner, or explosively clad steel options. Selection often is dictated by net wall thickness needed to handle service pressures and temperatures and related economics. Depending on pressure design code, higher-strength titanium alloys such as UNS R50550 (Grade 3), UNS R56320 (Grade 9), UNS R53400 (Grade 12), and UNS R56323 (Grade 28) may offer attractive, thinner-wall vessel construction options compared to lower-strength unalloyed grades or Ti-clad steel.

Thinner-wall piping and tubing in lower-strength, cold-formable titanium grades may be more cost-effective as seam-welded product forms, whereas seamless options may be more practical in heavy-wall tubular forms and/or those involving higher-strength titanium alloys. Also, the soft, highly cold-formable titanium UNS R50250 (Grade 1) and UNS R52250 (Grade 11) are highly reliable and erosion–corrosion-resistant, thin plate materials for lightweight, thermally efficient, seawater-cooled plate/frame and compact heat exchangers.

Titanium component surfaces can be coated with seawater-resistant coatings, polymeric coatings, or elastomeric sheathing, if needed for cosmetic, galvanic, or CP system protection purposes [13,18]. This can be accomplished by achieving an adequate anchor pattern (surface profile), and/or by precoating treatments to increase adhesion, such as chemical conversion coatings, anodised films, or special sol–gel primers.

Acknowledgements

The author acknowledges the support and commitment of RTI International Metals, Inc., in the preparation of this titanium chapter.

References

1. 'Titanium for energy and industrial applications', (ed. D. Eylon), The Metallurgical Society of AIME. AIME, Warrendale, PA, 1981.
2. J. A. Mountford, Jr and J. S. Grauman: International Workshop on 'Advanced Materials for Marine Construction', New Orleans, LA, 4–7 February 1997, (ed. G. R. Edwards *et al.*), 107–128; 1997, New York, NY, American Bureau of Shipping.
3. J. A. Mountford, Jr: International Workshop on 'Advanced Materials for Marine Construction', New Orleans, LA, 4–7 February 1997, (ed. G. R. Edwards *et al.*), 149–160; 1997, New York, NY, American Bureau of Shipping.
4. L. Lunde: 'Properties and applications of titanium and titanium alloys', 24–29; 1994, *Titanium Europe.*
5. 'Thinner wall welded titanium tubes for seawater desalination plants', Technical Brochure, Japan Titanium Society, Tokyo, Japan, 1984.
6. R. W. Schutz and M. R. Scaturro: UK Corrosion '88 with Eurocorr Conference Proceedings. Vol. 1, Published by the Institute of Corrosion Science and Technology, Birmingham, UK, 1988, pp. 285–300.

7. R. W. Schutz: 24th Annual Offshore Technology Conference, Houston, TX, 4–7 May 1992, OTC, Richardson, TX, Paper 6909.

8. W. L. Adamson and R. W. Schutz: *Naval Eng. J.*, 1987, **99**, (3), 124–134.

9. R. W. Schutz and M. R. Scaturro: *Navy Eng. J.*, 1991, **103**, (3), 175–191.

10. R. W. Schutz and M. R. Scaturro: *Sea Technol.*, 1988, **29**, (6), 49–54.

11. J. A. Mountford, Jr and M. R. Scaturro: 2009 SNAME Annual Meeting and Ship Production Symposium, The Society of Naval Architects and Marine Engineers, Vol. 2, 63; 2009, SNAME, Jersey City, NJ.

12. C. F. Baxter and R. W. Schutz: International Workshop on Advanced Materials for Marine Construction, New Orleans, LA, 4–7 February 1997, American Bureau of Shipping, Houston, TX, pp. 129–148.

13. C. F. Baxter, R. W. Schutz and C. S. Caldwell: 2007 Offshore Technology Conference, Houston, TX, 30 April–3 May 2007, OTC, Richardson, TX, Paper 18624.

14. L. Lunde et al.: Titanium '95: Science and Technology, Proc. Eighth World Conference on Titanium, Birmingham, UK, 1995, (ed. P. A. Blenkinsop, W. J. Evans, H. M. Flower), Vol. II, 1711–1718; 1996, London, UK, IOM[3].

15. L. Lunde and M. Seiersten: in 'Titanium and titanium alloys', (ed. C. Leyens, M. Peters), 483–497; 2003, Weinheim, Germany, Wiley-VCH GmbH & Co. KGaA.

16. 'Titanium: A technical guide', 2nd edn, (ed. M. J. Donachie); 2000, Materials Park, OH, ASM International.

17. 'Titanium and titanium alloys – Fundamentals and applications', (ed. C. Leyens, M. Peters); 2003, Weinheim, Germany, Wiley-VCH GmbH & Co., KGaA.

18. R. W. Schutz: CORROSION/2001, Houston, TX, 11–16 March 2001, NACE International, Houston, TX, Paper 010032001.

19. R. W. Schutz: 'Metals handbook', Vol. 13B, 252–299; 2005, Materials Park, OH, ASM International.

20. J. A. Beavers, G. H. Koch and W. E. Berry: 'Corrosion of metals in marine environments', MCIC-86-50, Metals and Ceramics Information Center Report, Chapter 3, MCIC, Columbus, OH, 1986.

21. J. B. Cotton and B. P. Downing: *Trans. Inst. Marine Eng.*, 1957, **69**, (8), 311–319.

22. D. K. Peacock: *Underwater Technol.*, 1996, **21**, (4), 23–30.

23. R. W. Schutz: *Mater. Perform.*, 1991, **30**, (1), 58.

24. T. S. Lee, E. W. Thiele and J. H. Waldorf: *Mater. Perform.*, 1984, **23**, (11), 44–46.

25. R. Holthe, E. Bardal and P. O. Gartland: *Mater. Perform.*, 1989, **28**, (6), 16–23.

26. D. M. Aylor, R. A. Hays and L. S. Marshall: CORROSION/2000, Orlando, FL, 26–31 March 2000, NACE International, Houston, TX, Paper 00640.

27. P. O. Gartland, F. Bjørnås and R. W. Schutz: CORROSION/97, New Orleans, LA, 10–14 March 1997, NACE International, Houston, TX, Paper 477.

28. H. Nagler, D. F. Hasson and C. S. Young: Metals handbook', 9th edn, Vol. 6, 446–456; 1983, Materials Park, OH, ASM International.

Roger Francis

Rolled Alloys, Unit 16, Walker Industrial Park, Walker Road, Blackburn BB1 2QE, UK

rfrancis@rolledalloys.com

In the past, galvanic corrosion has caused numerous failures of marine systems, some of them very costly. Galvanic corrosion is briefly mentioned in the previous chapters in this guide but this chapter gives more details on the causes of galvanic corrosion, the main factors affecting its severity, and some preventative measures that can be adopted for marine service conditions. Additional information can be found in reference 1.

8.1   Introduction

When a metal is immersed in an electrically conducting liquid, it takes up an electro-chemical potential (also known as the corrosion potential). This is determined by the equilibrium between the anodic and cathodic reactions occurring on the surface and it is usually measured with reference to a standard electrode, such as the saturated calomel electrode (SCE) or silver/silver chloride ($Ag/AgCl_2$).

Galvanic corrosion occurs when two metals with different corrosion potentials are in electrical contact while immersed in an electrically conducting, corrosive liquid. Because the metals have different corrosion potentials in the liquid, electrons flow from the anode (more electronegative) metal to the cathode (more electropositive) to equalise the electrochemical potentials. This acts to increase the corrosion on the anode. The additional corrosion is called galvanic corrosion, also referred to as bimetallic corrosion, dissimilar metal corrosion, or contact corrosion.

The effect of coupling two metals increases the corrosion rate of the anode and reduces, or even suppresses, corrosion of the cathode. Hence, coupling a component to a sacrificial anode can prevent corrosion and this is the principle of cathodic protection (CP).

The basic requirements necessary to cause galvanic corrosion are:

1. An electrolyte bridging the two metals that may not be aggressive to the individual metals when they are not coupled, and may be in the form of bulk volume of solution (e.g. seawater), a condensed film, or a damp solid such as soil, salt deposits, or corrosion products.
2. An electrical connection between the two metals. This usually arises from direct physical contact but it can also arise when electrical continuity is established between two metals, for example, by an insulation-coated conductor, by structural metal work, or electrical earthing (grounding) systems. It is not necessary for the metal junction to be immersed in the electrolyte.
3. A sufficient difference in corrosion potential between the two metals to provide a significant galvanic current.

4. A sustained cathodic reaction on the more noble (electropositive) of the two metals; in most practical situations, this is the reduction of dissolved oxygen (i.e. the electrochemical reduction of dissolved oxygen to hydroxyl ions).

8.2   Factors affecting galvanic corrosion

The value of the corrosion potential for any alloy in seawater can be changed by a variety of factors such as temperature, velocity, biocide treatment, and others. However, the relative ranking of the alloys remains largely unchanged by these factors. The galvanic or electrochemical series shows the typical potentials of alloys in natural seawater, and the difference between the most noble and the most active is ~2 V (Fig. 8.1).

Materials coupled to metals that have more electropositive potentials are the ones that may suffer galvanic corrosion. However, the magnitude of the potential difference alone is not sufficient to predict the extent of galvanic corrosion. For instance, metals with a potential difference of only 50 mV have shown severe galvanic corrosion problems, while other metals with a potential difference of 800 mV have been successfully coupled together. The potential difference provides no information on the kinetics of galvanic corrosion, which depends on the current flowing between the two metals.

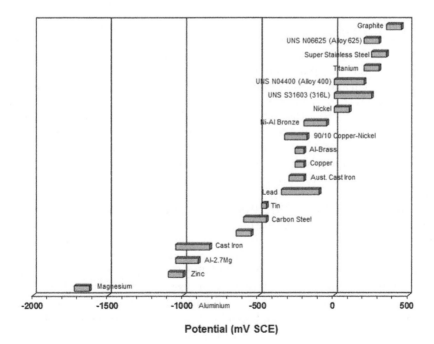

8.1 Galvanic series in natural seawater at 10 °C
(Data courtesy of R. Francis and SINTEF [Stiftelsen SINTEF, Box 4760 Sluppen, NO-7465 Trondheim, Norway])

The main factors that affect galvanic corrosion are:

- area ratio
- cathode efficiency
- electrolyte
- aeration and flow rate, and
- metallurgical condition and composition.

Temperature can affect the cathode efficiency and the degree of aeration.

8.2.1   Area ratio

The ratio of the exposed areas of the anode and cathode materials in a couple is very important in the consideration of the likelihood of galvanic corrosion. The larger the cathode compared with the anode, the more oxygen reduction, or other cathodic reaction, can occur; therefore, the greater the galvanic current. This results in more corrosion on the anode.

Adverse area ratios are likely to occur with fasteners and at joints. Weld, braze, or rivet metal should be of the same potential as, or better still, slightly cathodic to the base or parent metal. For example, the use of mild steel bolts or rivets to fasten nickel, copper, or stainless steel alloy plates should be avoided; however, joining carbon steel plates with copper rivets is acceptable.

The effect of area ratio is not confined to general corrosion, but can affect other types of corrosion. This is demonstrated by a failure that occurred on an offshore platform in the North Sea. The failure occurred at the junction of a 6% Mo stainless steel pipe and one made of 90/10 copper–nickel [2]. In chlorinated seawater, this would not normally result in any significant acceleration of the corrosion of the 90/10 alloy [1]. However, there was turbulence at the junction because of a protruding gasket and this, combined with the coupling to stainless steel, induced erosion–corrosion. This involved a small area and the cathode-to-anode area ratio was then very high, resulting in rapid penetration of the copper–nickel [2].

Extremely small anodic areas, such as cracks or pinholes, exist at discontinuities in cathodic coatings such as magnetite (mill scale on iron) and copper plating on steel. In seawater, where electrolyte conductivity is high, it is necessary to apply protection or, in the case of metal coatings, to specify a thickness of adequate integrity. Similar considerations apply to pores or defects in a coating if the coated metal is in contact with a more electropositive metal (i.e. the cathode-to-anode area ratio is dramatically increased compared with uncoated metal). This should be avoided if at all possible, because the rates of penetration are very high.

If the area ratio is selected correctly, dissimilar metals are frequently used together successfully. One example is the use of UNS J92900 (ASTM A351 [3], Grade CF8M, commonly known as Type 316) stainless steel pump impellers in austenitic cast iron pump casings in seawater. Another is the use of high-alloy valve trim (seats, stems, and others) in metals such as UNS N06625 (alloy 625) and Ni–Cu alloy UNS N05500 (alloy K-500), with gunmetal or aluminium bronze valve bodies in seawater. In both cases, the large area of the anode compared to the cathode means the increased corrosion is small and well within design limits for the valve or pump body.

8.2.2   Electrode efficiency

When a metal forms part of a galvanic couple, the rate of the anodic or cathodic reaction is not the same for all metals. This is because the kinetics of the reaction vary from metal to metal. The rate of the reaction at the metal surface determines its efficiency as an anode or cathode. In galvanic corrosion, a change of efficiency in the cathodic reaction usually has an effect more significant than one in the anodic reaction. The more efficient the cathodic reaction is, the greater the current and the greater the corrosion.

Some metals, such as titanium, are not very efficient at reducing dissolved oxygen compared, for example, with copper alloys. So it is possible for a metal, such as carbon steel, to corrode more when coupled to a copper alloy than to titanium, despite titanium being much more electropositive than copper. Titanium still accelerates the corrosion rate, but not as much as copper.

The cathodic efficiency can also change under different conditions. An example is the behaviour of high-alloy stainless steels, nickel–chromium–molybdenum alloys, or titanium and its alloys. In natural seawater, these alloys develop a biofilm after about 2 to 20 days of immersion. This biofilm makes these alloys very efficient cathodes [1].

When chlorine/hypochlorite is added to the seawater to control fouling (e.g. 1 mg L^{-1} at 20 °C), the biofilm does not form and the potential increases to about +600 mV SCE. The alloys are less efficient as cathodes because the cathodic reaction is now the reduction of hypochlorite to chloride ions [1]. In hot seawater, where the temperature is 25 to 30 °C above local ambient, the biofilm cannot form and these passive metals become very inefficient cathodes [1]. Therefore, galvanic corrosion is usually more severe in natural seawater compared with chlorinated seawater, or hot seawater. Note that local ambient temperature varies around the globe and with the seasons. In the Antarctic, it is about 0 °C, in the North Sea, it is about 6 °C, and off the North Carolina coast of the United States, for example, it can vary from ~10 °C in winter to ~30 °C in summer.

8.2.3   Hydrogen embrittlement

In some instances, the cathodic reaction can become hydrogen evolution. In seawater, this becomes significant at potentials of –850 mV SCE and more electronegative. This means that alloys that are susceptible to hydrogen embrittlement (HE) may embrittle or crack when connected to anodes that produce these electronegative potentials. Materials such as UNS N04400 (alloy 400) and UNS N05500 (alloy K-500), some cold-worked nickel alloys, duplex stainless steels, and high-strength carbon and low-alloy steels can all embrittle when coupled to some aluminium alloys, zinc, or magnesium. In addition, titanium can form hydrides, which make the metal brittle if coupled to these metals.

As an example, Efird [4] reported the failure of UNS N05500 (alloy K-500) bolts by HE, when used subsea and connected to zinc anodes that were installed to protect adjacent carbon steel. Properly applied CP can completely suppress galvanic corrosion, but may induce HE of susceptible alloys. See section 8.6.4 *Cathodic protection* later in this chapter.

8.3   Aeration and flow rate

The majority of practical situations involving galvanic corrosion develop in seawater under conditions in which the cathodic reaction is the reduction of dissolved oxygen. As with single metal corrosion, galvanic corrosion is, therefore, partly dependent on the rate at which oxygen can diffuse to the cathodic surface from the bulk electrolyte.

The galvanic corrosion rate of many copper- and nickel-based alloys, and of stainless steels in seawater, depends on the flow rate of the water as well as the area ratio. Copper and copper–nickel alloys tend to become more noble (i.e. more electropositive), and corrode less as the flow rate increases unless flow rates exceed the erosion–corrosion limit [5]. In well-aerated, flowing solutions, nickel-based alloys and stainless steels are likely to become passive and corrode less than under stagnant conditions.

Noble metals, such as platinum, silver, and copper, on which the naturally formed oxide films are very thin and easily reduced to metal, act as efficient cathodes and therefore tend to promote galvanic corrosion. However, stainless steel and titanium both have a stable oxide film and are poor cathodes. In flowing, aerated seawater, the oxide film is likely to thicken, thus further diminishing galvanic corrosion of the anodic metal. Graphite is also very electropositive (Fig. 8.1) in aerated, near-neutral solutions and it forms no oxide film. Therefore, like platinum, graphite is a very efficient cathode and, as a result of its widespread use in engineering applications (e.g. gaskets and composites), it has caused severe galvanic corrosion of many metals on a number of occasions.

The most common problem with graphite is graphite-loaded gaskets at flanged joints. Failures have occurred in both the British and Norwegian sectors of the North Sea with high-alloy stainless steel piping [6,7]. The solution was to use synthetic fibre gaskets.

In neutral electrolytes, complete deaeration, in many instances, suppresses single metal and bimetallic or galvanic corrosion. However, under such anaerobic conditions, corrosion can occur if sulphate-reducing bacteria (SRB) are present (i.e. an alternative reaction to the reduction in dissolved oxygen is available). In closed, recirculating systems, deaeration is very practical and very low dissolved oxygen concentrations can be maintained with chemical oxygen scavengers. Theoretically, the corrosion of iron and steel in a closed recirculating system should consume the dissolved oxygen fairly quickly. In practice, there are often places where small quantities of air can be drawn into the system, requiring the use of corrosion inhibitors and oxygen scavengers.

In once-through cooling systems, it is more difficult to control oxygen levels and 100% deaeration is rarely achieved. In land-based installations, such as multistage flash desalination plants, deaeration to <20 ppb is routinely achieved. (Note that, for practical purposes, 1000 ppb = 1 mg L^{-1}.) Under these conditions, the corrosion rate of carbon steel is very low. However, on offshore oil platforms, where space and weight are at a premium, deaeration of seawater for injection rarely gets below 50 ppb dissolved oxygen and levels of 100 or 200 ppb are not uncommon. Under these conditions, corrosion rates are significant for some metals, such as carbon steel, and galvanic corrosion is also a possibility.

As the temperature of seawater rises, the solubility of dissolved oxygen decreases, being about 7 mg L^{-1} at 15 °C, 4 mg L^{-1} at 50 °C, and about 1 mg L^{-1} at 90 °C. Consequently, the tendency for galvanic corrosion can decrease at high seawater temperatures.

8.3.1 Metallurgical condition and composition

In multiphase alloys, it is common for one phase to be anodic to another. Normally these differences are small and do not give rise to problems, but in some instances, severe galvanic corrosion can occur, such as when the corroding phase has a smaller exposed area than the cathodic phase.

A good example is Nickel Aluminium Bronze (NAB, UNS C95800), which can suffer corrosion when a particular phase (Kappa III) that is normally cathodic becomes anodic at crevices and shielded areas [8]. This corrosion can be exacerbated if the NAB is coupled to a more noble metal, such as a high-alloy stainless steel [9]. The selective phase attack is much less likely to occur if the alloy forms a protective film on the open surfaces in aerated, flowing seawater when first exposed. The problem can also be avoided using one of the methods discussed in section 8.6 *Methods of prevention* later in this chapter.

Another example of selective phase attack is cast iron in seawater, where the graphite particles are cathodic to the metal matrix. This is not a severe problem because the graphite area is so much smaller than that of the metal matrix, and it can take 15 years or more before sufficient graphite is exposed to significantly change this area ratio [1].

Local changes in composition can also arise at joints made by welding for most alloys, both in the weld bead and in the heat affected zone of the parent metal. Problems are usually avoided by selecting appropriate filler metals and welding techniques. Severe corrosion has occurred at welds in pipes where the welding operation produced a weld bead that was ~50 mV electronegative to the parent metal. The large area of the cathode resulted in rapid corrosion of the weld metal [1].

Galvanic corrosion can occur between alloys of similar types, but somewhat different composition. Thus, 90/10 copper–nickel is anodic to 70/30 copper–nickel, and can corrode more when coupled to a larger area of 70/30 copper–nickel. The 70/30 copper–nickel is the preferred filler metal when 90/10 copper–nickel is welded, as this creates a couple with a large anode and a small cathode. Austenitic stainless steels are often cathodic to martensitic stainless alloys; austenitic cast iron is more cathodic than other types of cast iron and therefore induces additional corrosion. Cast iron that suffers significant graphitic corrosion induces additional corrosion of noncorroded alloy, because the surface of the corroded cast iron is then essentially just graphite that is strongly electropositive to most metals, and it is also a very efficient cathode, as described earlier in this chapter.

Under some conditions, it is possible to get galvanic corrosion with a single metal. This occurs when part of a system is replaced with new metal while some of the old metal remains. If there is a potential difference between the filmed metal and the bare metal, significant galvanic corrosion may occur before the new metal forms a protective film. One solution is to clean the old metal to remove the film so all of the system starts at the same potential. Other cases in which galvanic corrosion can occur on a single metal are on stainless steels with heat tints (where there is significant chromium depletion beneath the tint), and on a metal where parts are exposed to aerated water and others to deaerated water.

8.4   Immersed conditions

When the performance of materials in seawater is reviewed, these can be divided into four major groups, according to their corrosion behaviour. Table 8.1 shows the four alloy groups and the common alloys found in each class. Group 1 includes all of the passive alloys that usually do not develop corrosion in seawater at ambient temperature.

Table 8.1 Alloy groupings for seawater service at ambient temperature

Group	Type	Alloy
1	Noble; passive; highly corrosion-resistant	Nickel–chromium–molybdenum alloys (Mo > 7%) 6% Mo austenitic stainless steel Superduplex stainless steel Titanium and its alloys
2	Passive; not fully corrosion-resistant	UNS N04400/UNS N05500; UNS N08904; UNS S31803/S32205 (22% Cr Duplex); UNS N08825; UNS N08020; UNS S31603
3	Moderate corrosion resistance	Copper alloys Austenitic cast iron
4	Poor corrosion resistance (anodic materials)	Carbon steel Cast iron
		Aluminium alloys

Some definitions are required to identify alloy classes. The nickel–chromium–molybdenum alloy class covers all of the nickel alloys with more than 7 wt% molybdenum, such as UNS N06625 (alloy 625) and UNS N10276 (alloy C-276). Titanium covers not only Grade 2 (commercially pure) but also the common commercial alloys such as Ti–6Al–4V and Ti–0.15% Pd. High-alloy austenitic stainless steel includes all those alloys with a molybdenum content equal to 6 wt% or greater and having a pitting resistance equivalent number (PREN) > 40 (for which PREN = %Cr + 3.3 (%Mo + 0.5 × %W) + 16 × %N; all values as wt%). The superduplex stainless steels include all duplex alloys with 25% or more chromium and a PREN > 40.

In Group 2 are the passive alloys that are not totally immune to localised corrosion (e.g. crevice corrosion and pitting corrosion) in seawater. Such alloys are mostly stainless steels and iron–nickel–chromium–molybdenum alloys with PREN values <40 (e.g. UNS S31603 [alloy 316L] stainless steel and UNS N08825 [alloy 825]). Another type of alloys that fall into this group are the high-nickel–copper alloys, the best known of which are UNS N04400 (alloy 400) and UNS N05500 (alloy K-500). These can suffer crevice corrosion in seawater [10,11]. The common Group 2 alloys are shown in Table 8.1.

Group 3, also shown in Table 8.1, includes alloys of moderate corrosion resistance that do not show true passivity. The most commonly used in seawater are the copper alloys (brasses, bronzes, and copper–nickels), and the austenitic cast irons. These cast irons are more noble than conventional cast irons because of their high nickel content.

Group 4 covers alloys with poor corrosion resistance and is subdivided into irons and steels, and aluminium alloys. This is because there can be a substantial potential difference between aluminium alloys and iron and steel in seawater, so they are treated separately as shown in Table 8.1.

The principal reason for grouping the alloys as shown in Table 8.1 is that there is ample evidence that bimetallic couples between any of the alloys within a particular group do not generally lead to any significant galvanic corrosion [1]. This does not necessarily mean that no corrosion will occur, but no significant additional corrosion occurs. For example, naval brass can dezincify in seawater, but coupling it to 90/10 copper–nickel does not significantly increase the rate of penetration.

There are a few exceptions to this rule, but they are not common, and even then, with a suitable area ratio, galvanic corrosion is not a problem. Therefore, when there is any doubt about the compatibility of materials, the selection of an alloy from the same group generally prevents problems. There are clearly cases in which alloys from different groups can be connected successfully, but selecting alloys from the same group is a good rule of thumb when in doubt.

8.5   Atmospheric corrosion

Atmospheric corrosion is somewhat different from full immersion in seawater and this influences galvanic corrosion as well. There are many industrial plants and pieces of equipment designed to operate in marine atmospheres and they must be protected from corrosion. While coatings are often used, some equipment must remain uncoated and the selection of the correct materials is important to avoid galvanic corrosion.

The most important factor is water. Corrosion in direct contact with air at ambient temperature above its dew point is generally not significant from an engineering perspective. Water must condense on the exposed metal surfaces or precipitate as rain. This is followed by evaporation as the components dry out; thus, a most important parameter for atmospheric corrosion is the time of wetness. Condensation is controlled by the relative humidity and it is generally accepted that a relative humidity of 80% or more, plus a temperature greater than 0 °C, are required for significant condensation [12]. Rainfall is dependent on the local weather and can vary greatly.

Corrosion in the atmosphere is generally not as great as that when the metal is fully immersed in water. However, galvanic corrosion rates can be as high as those under immersed conditions.

As the water evaporates, anything dissolved in it concentrates, and when these species are aggressive to a metal, the corrosion rate increases as the water evaporates. The common species found in condensation and rainwater are briefly discussed below.

1. Oxygen – this is always present in the atmosphere and when the water films are thin, rapid diffusion of oxygen can occur. As a result, there is always a ready supply of oxygen to sustain the cathodic reaction.
2. Carbon dioxide – this gas is common in the atmosphere and dissolves readily causing the water to be mildly acidic. It is not thought to be a major contributor to atmospheric corrosion [13].
3. Ammonia – this is common in rural areas and is readily soluble causing increases in pH. This is beneficial for most alloys, although stress corrosion cracking (SCC) of some copper alloys is possible.
4. Sulphur and nitrogen oxides – these are common in industrial and urban areas as effluent gases, such as fossil-fuelled power stations. All of these gases dissolve readily, forming strong acids that are the cause of acid rain.

In addition to these gases, there are two other factors that affect atmospheric corrosion: temperature and chlorides. Temperature is self-explanatory: the warmer the climate, the greater the rate of chemical reactions, including corrosion. Chloride is found adjacent to the sea and the closer a site is to the sea, the higher the chloride concentration in condensed water. This is demonstrated over the past 50 years by exposures at the LaQue Laboratory in North Carolina, USA, where corrosion rates of test panels on the lot 25 m from the sea are higher than those on the lot 250 m from the sea.

Using all these factors, sites are generally divided into four types by atmosphere: rural, urban, industrial, and marine (ISO 9226) [14]. Rural has low chloride and sulphur dioxide and may include ammonia. The urban atmosphere has some sulphur dioxide, while industrial areas have high sulphur dioxide and nitrogen oxide contents. Finally, the marine atmosphere is characterised by high chloride contents.

The cathodic reaction in atmospheric corrosion is usually the reduction of dissolved oxygen and, as described above, the reaction rates can be high, especially with thin water films. If the dissolved solids and gas content of the water is low, the conductivity is low, reducing the rate of corrosion, but as the films evaporate, the conductivity increases.

Galvanic corrosion in the atmosphere tends to be confined to the dissimilar metal joint where condensation has occurred. Therefore, area ratios are usually around 1:1. The corrosion rate varies as the species dissolved in the water vary, and as the water films evaporate and condense again. This is not readily simulated in the laboratory, but there is a body of data from long-term exposure trials of galvanic couples that is very useful.

Kucera and Mattsson published galvanic corrosion data for a variety of atmospheres [15]. In addition, the authors made an attempt to categorise the likelihood of galvanic corrosion for various bimetallic couples. Table 8.2 shows the atmospheric corrosion rates for 10 commonly used alloys. Weathering steel is a low-alloy carbon steel with ~0.5% copper. This produces an adherent rust layer that can lower atmospheric corrosion rates compared to carbon steel. In addition, aluminium was tested in the anodised condition, as this is commonly used to reduce aluminium corrosion. Care should be exercised when using the values in Table 8.2 as air quality in many industrial countries has improved since the 1970s, particularly with regard to sulphur dioxide.

Kucera and Mattsson tested many, but not all, of these metals in combination, and the results for marine atmospheres are summarised in Table 8.3 [15]. The goal was to show couples with no significant change in corrosion rate from uncoupled, a slight

Table 8.2 Corrosion of some common metals in various atmospheric environments [15]

Alloy	Corrosion rate (µm year^{-1})		
	Rural	Urban	Marine
Magnesium	6.5	11	14
Aluminium	0.1	0.5	0.6
Anodised aluminium	0	0.2	0.2
Zinc	0.5	2.5	1.3
Carbon steel	22	53	36
Weathering steel	22	48	32
Tin	1.4	2.7	10
Lead	0.3	0	0
Copper	1.2	1.4	3.2
Nickel	0.4	2.7	1.0
Stainless steel UNS S30400 (Type 304)	0.03	0.04	0.04

Table 8.3 Likelihood of galvanic corrosion in a marine atmosphere for some common metals [15]

Corroding metal[a]	Coupled metal[a]										
	Aluminium	Anodised Al	Zinc	Carbon steel	Weathering steel	Tin	Lead	Copper	Nickel	Stainless steel[a]	Chromium
Magnesium	░	░	░	■		▨	■	■	■	■	▨
Aluminium	░	░	░	■		▨	■	■	▨	▨	▨
Anodised aluminium		░		■		▨	▨	■	▨	▨	░
Zinc	░	▨	░	■		░	■	■	░	▨	▨
Carbon steel	░	▨	░	░	░	▨	░	▨	▨	▨	░
Weathering steel	░			░	░	░	░	░	░	▨	
Tin								▨	▨	░	
Lead	░	░	░	■		▨	░	▨	▨	░	■
Copper	░	░	░	░	░	░	░	░	░	░	░
Nickel	░	░	░	░	░	░	░	░	░	▨	▨
Stainless steel[a]											

[a]Note that titanium and nickel alloys behave like stainless steel.

░ No likelihood of galvanic corrosion   ▨ Some increase in corrosion

■ Large increase in corrosion   ☐ No data

increase in corrosion rate, and a greater increase in corrosion rate. This has to be compared with the uncoupled corrosion rate in Table 8.2, as an increase in corrosion rate may still not represent much of a loss of metal and may be acceptable for a particular application.

There are no absolutes here and some metal combinations may be acceptable in one case and not in another depending on the criticality of the application. Chromium is included in Table 8.3 because of its widespread use as electroplating in atmospheric applications.

Note that in tropical marine atmospheres, galvanic corrosion can be more severe than in temperate climates [1].

8.6   Methods of prevention

The following sections review the main methods of preventing and combating galvanic corrosion. To do this properly, it is necessary to know as much as possible about the seawater conditions. This includes not only salinity and temperature, but also any additions (such as biocides) and dissolved oxygen content. The more information that is available, the more confidence can be placed in the decisions made to mitigate galvanic corrosion.

8.6.1   Materials

When the design is being completed and before any materials are purchased, there is the maximum opportunity to select galvanically compatible metals for the specific application. The following guidelines offer some helpful suggestions.

- Use compatible metals wherever possible. Try to select alloys from the same group, as shown in Table 8.1. Avoid metals that are far apart in the galvanic series (Fig. 8.1), unless it is well documented that they can be coupled successfully (e.g. 90/10 copper–nickel and high-alloy stainless steel) in chlorinated seawater.
- When welding, pay particular attention to the condition of the weld after fabrication. It should be cathodic to the parent metal. When carrying out autogenous welding, choose welding parameters that achieve this. When using filler metal, select one that produces weld metal with a similar potential or is electropositive to the parent metal. With such a high cathode-to-anode area ratio as occurs in a typical welded joint, a weld metal potential of only 15 mV negative to that of the surrounding parent metal can result in severe weld metal corrosion in some fluids [1].
- When compatible materials cannot be used for whatever reason, one of the following methods of prevention of galvanic corrosion should be considered:
  - isolation and separation
  - coatings
  - cathodic protection
  - inhibitors, or
  - design.

8.6.2   Isolation and separation

If the anode and cathode can be prevented from coming into electrical contact, then no galvanic corrosion will occur. In some situations this can be done quite conveniently, while in others it is impractical.

Several methods for achieving this when bolting together two flat surfaces were discussed by Parkins and Chandler, depending on whether it is the bolts or one of the plates that is the anode [1,16].

Another method of preventing galvanic corrosion between two adjacent metals could involve the use of a thick coat of nonconducting, insoluble mastic at a joint. The mastic may additionally contain chemicals to inhibit corrosion of the metals as they are leached out.

One method for preventing galvanic corrosion of piping at dissimilar metal joints is the isolating flange. This method uses a nonconducting insert between the two metal flanges and nonconducting sleeves and washers for the bolts. This means the two metals are no longer in electrical contact; however, there are two major disadvantages to this technique. First, the sleeves and washers must be undamaged and correctly assembled for the isolation flange to work properly. On many industrial sites, sufficiently skilled labour is not available and this is difficult to ensure.

Second, there is frequently a requirement by the electrical engineers to earth (ground) all major items of metal equipment for safety reasons. This has the effect of shorting across the isolation flange, and, unfortunately, it is a common industrial practice [17]. Pipe hangers or supports also act as inadvertent electrical shorting between adjacent pipe runs. When isolating flanges are used, it is important that the resistance between the two components be measured before start-up to ensure electrical isolation is achieved.

An alternative method of separating the anode and cathode when one is a pipe and the other is a pipe, vessel, valve, etc., is to use an insulating spool piece. This is a length of pipe between the two items that is either internally coated or is made from a nonmetallic material, for example, fibre-reinforced plastic/glass-reinforced plastic (FRP/GRP). Note that the coating must extend onto the flange faces as well as along the pipe bore.

The metal pipe being coated must be made of the same metal as that forming the cathode or a compatible corrosion-resistant alloy, so defects in the coating do not suffer rapid localised corrosion. If a coating is being used, it should have a life consistent with the planned maintenance intervals for the plant. A coating that needs replacing at more frequent intervals is often not cost-effective, and an alternative solution to the galvanic corrosion problem may be required.

The length of the insulating spool piece depends on the electrochemical properties of the two metals and the particular fluid under consideration. An example for seawater serves to demonstrate this principle. When a copper alloy, such as 90/10 copper–nickel, is joined to a Group 1 alloy, it is common to use an insulating spool piece approximately six pipe diameters long, provided there is no undue turbulence to cause localised erosion–corrosion of the copper alloy.

For example, an oil tanker was fitted with 90/10 copper–nickel piping and a titanium plate heat exchanger. Galvanic corrosion occurred on the copper–nickel adjacent to the heat exchanger, and the problem was solved by fitting a coated titanium spool piece five diameters long. The length of the spool was limited by the available space in the ship, but it prevented further excessive corrosion of the copper–nickel pipe.

Bardal et al., showed that with high-alloy stainless steel and 90/10 copper–nickel in chlorinated seawater, an insulating spool piece is unnecessary [2]. However, if the chlorination were likely to be turned off at all, an insulating spool piece would be a wise precaution to prevent corrosion at the dissimilar metal joint. In natural seawater, the corrosion rate of 90/10 copper–nickel was ~1.2 mm year^{-1} at the junction with a stainless steel pipe.

Gartland and Drugli examined the use of coated spool pieces between carbon steel and high-alloy stainless steel [18]. They found that an insulating spool piece 2 to 2.5 m long was required for most pipe size combinations carrying aerated seawater.

8.6.3 Coatings

It is sometimes possible to use coatings on the affected area to prevent a galvanic corrosion problem. Preferably *both* the anode and cathode should be coated, but it is often sufficient just to coat the cathode. A small defect in the coating still means that only a very tiny cathode area is exposed to the seawater. In no case should only the anode be coated unless CP is also used. The cathode-to-anode area ratio is very high at defects in the coating, and the rate of corrosion at the anode usually increases dramatically.

It is important to select coatings that have a life similar to that of the components they are protecting, especially if access is limited or difficult. Otherwise, regular maintenance of the coating is necessary to prevent galvanic corrosion as the coating breaks down.

An alternative method to organic coatings is metallic coatings. The most commonly applied metals are aluminium and zinc, although many other metals can be used as coatings. An example is the coupling of an aluminium alloy and carbon steel in seawater, which accelerates the corrosion of the aluminium alloy. The steel could be sprayed with zinc or aluminium to prevent galvanic corrosion.

The most common, and cheapest, method is thermal spraying. However, this produces porous coatings and, for maximum life, polymeric sealers (e.g. silicone-based) are required after coating. An alternative is to use high-velocity oxy fuel (HVOF) spraying that produces very dense coatings with no need for sealing, although this process is a little more costly.

Note that where a metallic coating is porous and no sealer is used, it is essential that the coating is anodic to the base metal; otherwise accelerated corrosion occurs at the defects in the coating because of the high cathode-to-anode area ratio. An example of a poor choice of coating is chromium plating on steel, which is unsuitable in seawater.

In the case of atmospheric corrosion only, there are some additional coating measures that can be considered. These rely on the galvanic corrosion being confined to the joint area with an anode-to-cathode area ratio of approximately 1:1. One solution is to use a non-conducting mastic or polymeric sealer at the junction, which effectively increases the resistance between the anode and cathode. A similar effect can be obtained with polymeric tapes that may contain inhibitors that can leach out. The tape is usually applied on the jointed faces and up to about 25 mm to either side of the bimetallic joint. Rubber coating is also used to successfully prevent atmospheric galvanic corrosion.

8.6.4 Cathodic protection

This method of protection relies on changing the coupled potential of two dissimilar metals. It is not always necessary to suppress corrosion totally, but only to prevent or reduce the corrosion caused by the galvanic couple.

Two methods are used: impressed current and sacrificial (galvanic) anodes. With the first, a current sufficient to change the potential to the desired value is applied. With a sacrificial anode, the final potential is usually close to the open-circuit (corrosion) potential of the anode, so the metal chosen for the anodes is important. Not only must the potential be in the desired range, but the anode must also be of a size and have a life and a cost that are practical.

Aluminium alloy and zinc anodes are commonly used but sometimes the potential of these anodes is too negative (~ –1 V SCE) and they can cause hydrogen problems or excessive scaling. The solution then is to use anodes with a more electropositive potential, such as mild steel (C < 0.1%). These are sufficiently negative to prevent corrosion, such as dezincification of duplex brasses, and they also prevent hydriding of titanium heat-exchanger tubes because hydrogen is not produced [1]. Mild steel anodes also prevent pitting and/or crevice corrosion of low-alloy stainless steels, such as UNS S31603 (alloy 316L), in seawater [19].

An alternative to mild steel is Al–Ga that has a potential of about –850 mV SCE compared with ~ –1040 mV SCE for zinc and normal sacrificial aluminium anodes, and ~ –500 mV SCE for mild steel. Al–Ga anodes have found some applications in which their intermediate potential is useful and these do not produce a significant amount of hydrogen [20,21].

The same principle can also be applied in piping systems by fitting a sacrificial spool piece, usually of carbon steel, five to six pipe diameters long between the two components of the bimetallic couple. In seawater, the spool piece requires replacement every 1 to 2 years, but if access is easy, this can be a cost-effective solution. Remember that iron corrosion products from the spool piece will be carried downstream and this must be acceptable in the rest of the system. The corrosion products usually form soft deposits of hydrated iron oxides up to ~1 mm thick, which may not always be desirable.

8.6.5 Inhibitors

Inhibitors are rarely used in once-through seawater systems, although they may be used in recirculating closed systems, or in seawater injection systems in the oil and gas industry. Note that ferrous sulphate, sometimes added to improve the corrosion resistance of copper alloy heat-exchanger tubes, is not an inhibitor in the classical sense. The iron hydroxide (FeOOH) deposit that is produced acts as a barrier between the turbulent water and the protective film on the metal surface [22].

When inhibitors are used, it is important to select one that is compatible with all of the metals in the system.

The use of mastics and tapes to prevent atmospheric galvanic corrosion was discussed in section 8.6.3 *Coatings* earlier in this chapter. In some cases, these can be obtained with inhibitors to provide additional protection against galvanic corrosion. As with fully immersed conditions, it is important that suitable inhibitors are used that are compatible with both metals in the couple being protected.

A slightly different method of inhibition is deaeration, which usually can only be applied to recirculating systems. If hydrogen evolution is not the cathodic reaction, the addition of oxygen scavengers to prevent the cathodic reduction of dissolved oxygen is a viable method of preventing corrosion, including galvanic corrosion. It is essential to prevent additional entry of oxygen into the system, or regular dosing of scavengers is necessary.

8.6.6 Design

Galvanic corrosion can be prevented by one of the methods described above if it is addressed early enough in the design stage. Once construction is started, changes and retrofits become more difficult and expensive.

There is a range of good practices that should be adopted to minimise the likelihood of galvanic corrosion. For structures exposed to the atmosphere, one of these is to prevent water accumulation and the entrapment of debris. Designs that allow water to run off freely pose fewer problems than those that create areas where water can collect.

Whenever possible, for fully immersed conditions, a design should be chosen to minimise the area of the cathode and/or maximise the area of the anode if galvanically incompatible metals are selected. If the cathode-to-anode area ratio is kept sufficiently small, additional methods of restricting galvanic corrosion may not be necessary.

Acknowledgements

Thanks are given to Anne-Marie Grolleau and Ron Schutz for their contributions to writing this chapter.

References

1. R. Francis: 'Galvanic corrosion – A practical guide for engineers'; 2001, Houston, TX, NACE International.
2. E. Bardal, R. Johnsen and P. O. Gartland: *Corrosion* 1984, 40, 12.
3. ASTM: 'Standard specification for castings, austenitic, for pressure-containing parts', ASTM A351/A351M (latest revision), ASTM, West Conshohocken, PA.
4. K. D. Efird: *Mater. Perform.*, 1985, **24**, (4), 37.
5. T. S. Lee, E. W. Thiele and J. H. Waldorf: *Mater. Perform.*, 1984, **23**, (11), 44.
6. O. Strandmyr and O. Hagerup: CORROSION/98, San Diego, CA, 22–27 March 1998, NACE International, Houston, TX, Paper 707.
7. S. Shrive: 12th Int. Symp. on 'Corrosion and Materials Offshore', Stavanger, Norway, January 1999, The Norwegian Society of Engineers and Technologists [NITO], Norges Ingeniør- og Teknologorganisasjon, 1999.
8. J. Rowlands: Proc. 8th Int. Congr. on 'Marine Corrosion and Fouling', Comité International Permanent pour la Recherche sur la Préservation des Matériaux en Milieu Marin Vol. 2, 1346–1351, COIPM, Newcastle, UK, 1981.
9. R. Francis: *Br. Corros. J.*, 1989, **34**, (2), 139.
10. E. B. Shone, R. E. Malpas and P. Gallagher: *Trans. Inst. Mar. Eng.*, 1998, **100**, 193.
11. R. Francis: 'The corrosion of copper and its alloys – A practical guide for engineers'; 2010, Houston, TX, NACE.
12. E. Mattsson: 'Basic corrosion technology for scientists and engineers', 2nd edn, 80; 1996, York, UK, The Institute of Materials, Minerals, and Mining.
13. L. L. Shreir, R. A. Jarman and G. T. Burstein: 'Corrosion', 3rd edn, 2:33; 1994, London, Butterworth Heinemann.
14. ISO: 'Corrosion of metals and alloys – Corrosivity of atmospheres – Determination of corrosion rate of standard specimens for the evaluation of corrosivity', ISO 9226 (latest revision), ISO, Geneva, Switzerland.
15. V. Kucera and E. Mattsson: in 'Atmospheric corrosion', (ed. W. H. Ailor), 561–574; 1982, New York, NY, John Wiley & Sons.
16. R. Parkins and M. Chandler: 'Corrosion control in engineering design'; 1978, London, UK, DOI/HMSO.
17. R. Johnsen and S. Olsen: CORROSION/92, Nashville, TN, 26 April–1 May 1992, NACE International, Houston, TX, Paper 397.
18. P. O. Gartland and J. M. Drugli: CORROSION/92, Nashville, TN, 26 April–1 May 1992, NACE International, Houston, TX, Paper 408.

19. T. S. Lee and A. H. Tuthill: *Mater. Perform.*, 1983, **22**, (1), 48.
20. J. P. Pautasso, H. Le Guyader and V. Debout: CORROSION/98, San Diego, CA, 22–27 March 1998, NACE International, Houston, TX, Paper 725.
21. E. Lemieux, K. E. Lucas, E. A. Hogan and A.-M. Grolleau: CORROSION/2002, Denver, CO, 7–11 April 2002, NACE International, Houston, TX, Paper 16.
22. J. E. Castle and M. Parvisi: Int. Colloquium on 'Choice of Materials for Condenser Tubes and Plates and Tube and Tightness Testing', French Nuclear Energy Society (SFEN), Avignon, France, 1982, 171.

Printed and bound by CPI Group (UK) Ltd, Croydon, CR0 4YY

23/10/2024

01777678-0011